SHIFTING BORDERLINES

SHIFTING BORDERLINES

HOW SCIENCE FICTION IS BECOMING SCIENCE

HAMMAD AZZAM

www.shiftingborderlines.com

Shifting Borderlines
Copyright © 2010 by Hammad Azzam.

All rights reserved.
Printed in the United States of America.

No part of this book may be used or reproduced in any manner whatsoever without written permission except in the case of brief quotations embodied in critical articles and reviews.

ISBN 9781-451-565249

First Edition

1. Information Technology. 2. Robotics. 3. Artificial Intelligence 4. Biotechnology

www.shiftingborderlines.com

to lina and her genes

CONTENTS

CHAPTER 1	Introduction	1
CHAPTER 2	Science Eccentricities	15
CHAPTER 3	Robotics	53
CHAPTER 4	Bio-Universe	89
CHAPTER 5	The Future	116
ENDNOTE		158

ACKNOWLEDGEMENTS	162
REFERENCES	165
ABOUT THE AUTHOR	174
INDEX	176

SHIFTING
BORDERLINES
HOW SCIENCE FICTION IS BECOMING SCIENCE

1

INTRODUCTION

It was your run of the mill end of a relatively undistinguished day at the office, filled with dull chores that were artfully and systematically glorified by the executive machine and spun as exotic careers for the working class. To hear the description of such prosaic professions coming in the form overhyped project designations like *Process Automation and System Aggregation Engines*, you would think that we were just short of deciphering the meaning of life.

Unfortunately, no deciphering was planned for that day. I dragged myself into the house and hugged the rocking chair for a few minutes looking for peace. I was pretty much drained and in need of unwinding before the eight hours rejuvenating sleep. One look at my wife and I realized that she was opting for the same lazy route. After a quiet dinner, we auto-piloted to the living room and the remote control ended up in her hand. It just happened to be two inches closer to her side of the couch, and that night the lazy gene had profoundly dominated the system, so I succumbed.

I rested my head on the couch pillow and watched the parade of channels march and eventually land on a Discovery Channel documentary on the International Space Station (ISS). I, probably like most people out there, was not aware what ISS was all about. A news flash about the latest successful shuttle launch gets as much reaction from me as the weather report in upstate New York. This documentary, however, was an eye opener. It made me realize that ISS is a grand ambition; a world collaboration to colonize the universe. It is actually a space station; a fixed point in space that people can travel to and come back from, and for the past twelve years we have been traveling up there and constructing, inch by inch, a structure that could eventually look like Deep Space Nine.

We are slowly but inevitably drifting in the Star Trek direction. It might be a long time before we get to *warp* speeds and journey through wormholes, but many of the features that seemed so far-fetched only twenty years ago are coming true right now. And many more are right around the corner. The borderlines between science and science fiction are quietly shifting. What was wild imagination last century is reality now. This has been happening slowly for decades, if not hundreds of years, but has taken on an accelerated form lately, so much so that we can actually perceive it clearly. A hundred years ago, flying a colossal metallic object was science fiction; fifty years later, planes filled the skies in such large numbers to warrant creating a new career category, air traffic controller; nowadays, airport traffic is measured by the number of planes landing per minute. Aviation took a century to become a commonplace channel of

Introduction

transportation; too long to be noticeable. The Internet, on the other hand, swept through the world in less than thirty years; and the cell phone growth story was even faster, rising from zero to a hundred percent penetration in countries like England in less than two decades.

This trend continued steady in the twenty first century. In fact, the past decade has been the most *noticeable* when it comes to advancements in science. New concepts, like the iPhone, have moved from blueprints into successful mass production with tens of millions of sets sold, all in less than 5 years. In fact, the iPhone has become a common item faster than my spellchecker had the chance to recognize the word (I can see the red squiggly lines under the word which you can't see in the printed manuscript). In the biological sciences field, consider the Human Genome Project, one of the most impressive scientific feats of humanity. It was essentially a large-scale global collaboration program, costing north of $2.7 billion to complete over a stretch of thirteen years. The same effort will soon be reproduce-able, custom-to-order, for less than $1000 and available within a few days. Now if you were one of those prominent scientists who, throughout the nineties, was leading one of the major biotechnology labs decoding genes for the human genome project, and you were reflecting on your past accomplishments, you would concede that the whole effort run by your team (dozens of highly specialized technicians and biologists) is now potentially achievable by one person in a lab in one day!

The Perfection of Science, or Not

The enabler of these quantum evolutionary leaps is the rapid progression of science and the proliferation of scientific fields tackling very specialized subjects like bioinformatics, artificial intelligence and nanotechnology. Science is growing fast and unabated. Wars, economic downturns, and all other detrimental events have little effect on the expansion of science. What started with few fields (mathematics, chemistry, biology) has mushroomed into thousands of domains that span every aspect of humanity. New disciplines like Very Large Scale Integration (VLSI) are commanding armies of loyal followers that diligently strive to create the next substantial breakthrough in the field.

In principle, science is the study of nature. It explains everything in the cosmos, from the nano-miniscule to the peta-colossal. It is universally acknowledged to be the most definitive and deterministic instrument available to humanity. At least that's what most of us think. And we see the attestations all around us, from fast moving cars, to high flying tons of steel; from incredibly realistic virtual-reality experiences, to exceedingly destructive weapons. Effectively, science is becoming a religion with a fast-growing number of followers who blindly believe in everything scientists shell out. Most of us do not understand the fundamental principles behind the theory, but we take for granted the products as they are handed down to us by the authoritatively intellectual. These principles are like millions of commandments that people accept and follow faithfully. For most of us science seems complete, thorough and precise; in other words it is perfect.

However, the truth, which is what science looks for in the first place, is that science is in no way comprehensively exact. Science exists because individuals (and as of lately groups) develop sincere interest in creating models that describe the physical world around us. These models aspire to consistently explain the behavior of physical phenomena in the form of

Introduction

axioms and theories which are the essential foundation of science. Some of these theories pick up steam and get acknowledged fairly quickly; others are ignored (and in some cases proven incorrect). Some last a short time before they are challenged or morphed into improved structures; others last centuries in their original pristine forms. Some thoroughly describe the phenomenon engaged, while others make assumptions that only superficially describe the elementary principles. Let's take an example one of the oldest and most understood set of laws known to us, gravity. Newton -- very brilliantly mind you-- explained how two objects attract each other with a force inversely proportional to the square of the distance between them. That is why Earth pulls you downward with a force equivalent to: *all of atoms in your body interacting with all the atoms of Earth*. Since Earth is below you, the force goes downward (actually down is the direction defined by the Earth's pull in the first place). With Newton's laws we have an understanding of how gravity functions – we have a theory. But what actually transpires between two objects to make them attract each other? Does Earth extend some form of invisible hands to the moon and the moon does the same to Earth and do the arms meet half way and bond? Does every atom on the moon know about every atom on Earth and have the ability to reach out to it to create the tides that we experience every day? Does this happen instantaneously? If we bring another moon in close proximity to Earth at very high speed, high enough to make it look like it popped in out of nowhere, how long before Earth will know about the new moon and feel the effect of its gravity? A few seconds? A few minutes? Instantly? Although there are attempts to explain how this happens -- through exchanges of, presumably, streams of a very small particle called graviton-- we are far from having a definite answer. In essence, the laws of gravity are just an aggregate general model that describe the result of an exchange between two objects, but does not necessarily explain the phenomenon.

Science limitations, however, were most severely manifested in the twentieth century. Traditionally scientists put together models to explain physical behavior. These models must adhere to basic axioms which are commonsense ground rules that form the groundwork for all science. Example of an axiom: *Two parallel lines cannot meet* or *Time is an absolute reference point*. Some of these axioms have held true through time, others have not been so lucky. Starting in 1905, with Einstein charging the squad, scientists mounted an assault on so many firmly grounded axioms and theories; an assault that rocked the foundation of physics. Why, you ask? After exhausting all commonsense methodologies in attempting to explain stubborn unexplainable phenomena, scientists started to realize that there are things in the universe that are not necessarily comprehensible by the feeble human mind. Things that our brains are not wired to grasp. What followed was Einstein's 1905 Relativity, and other convoluted theories that, to this date, are not only hard to follow by the common man, but are difficult to take in by the most elite in the scientific community. Pretty much no one can explain why some of these theories work, but everyone accepts that they do.

For example, Einstein's theory of Relativity suggests that the speed of light stays constant in every frame of reference. What does this mean? Well, it means that when you are standing still, and a ray of light is shone next to you from someone else standing still, it will zoom by you at the speed of light. This is rational. However, it also means that, if that same someone, standing still, shines a light next to you, while you are speeding at great velocity, that light is going to fly next to you at the

Introduction

speed of light, as if you are still standing still! This is utterly irrational. The consequence? One variable has to give (in order for the speed of light to stay constant in all frames of reference) because we know that speed is proportional to distance and inversely proportional to time. And when the speed of light stays constant in both frames of reference, something weird is going to happen; both distance and time will be affected. Without going into details, distances will shrink or expand and time will move *slower or faster*. And when Einstein says that distances will shrink, he means that the same stick that is measured by someone still to be one meter long will be measured by someone moving to be longer. And when Einstein says that time will move slower, he means that time will pass faster for the stand still person than it will for the person moving. All of this follows from basic mathematics and, once accepted, makes way for Spielberg to make his time-machine movies.

The whole problem initially cropped up when Michelson and Morley, in 1861, tried to measure the effect of the motion of Earth on the speed of light. Knowing that Earth moves at the incredibly high speed of 1600 km/hour, they attempted to measure the difference in the speeds of rays of light traveling parallel and perpendicular to the direction of the motion of Earth. But they could not find that difference. Every time they measured the speed of these rays (traveling in any direction, with, against, or perpendicular to the direction of motion of Earth) it came out to be exactly the same, 299,792,458 meters/sec. As hard as they tried (over months, at different times of the year, with Earth in different locations in its orbit around the sun) they could not detect any variations. Then Einstein came along and brought with him a model that explained this *constancy*. Differences were not to be found simply because the speed of light is *constant* in every framework. Einstein opted to take this almost supernatural contradiction in and see if we could live with it. And we did.

After Einstein ignited the science world with his daring propositions, the ball rolled fast[1]. The twentieth century ushered in an age where theories created by scientist throughout history, ranging from mediocre to brilliant, were extended, modified, and sometimes found to be wrong all together. What has become clear is that science is by no means final. Concepts are ever-changing and our views of how things work, especially on the extremes, are not engraved in stone. Less than 150 years ago atoms were called atoms because they were thought to be the smallest substances in the universe. 50 years later we started talking about protons, electrons and neutrons being the elementary ingredients of atoms. Protons and neutrons counts inside the atom defined the nature of matter and therefore defined the elements, which in turn created everything around us. Another 50 years later and we came to discover that even protons and neutrons were made of smaller particles called quarks. Nowadays theories are spawning that propose more elementary entities called strings that make up the quarks, that make up protons and neutrons, which in turn make up the atoms that were only 150 years earlier thought to be the fundamental ingredients of the world.

Boundary Conditions

Science is limited by our reach and, for the longest time, our reach has been constrained by our senses. Take cosmology, the

---

[1] Ironically, the next set of strange theories to bombard physics did not go down well with Einstein. He spent the rest of his life unsuccessfully fighting the weirdness that quantum theory brought to the world of physics.

Introduction

study of the universe at the macro level, for instance. The nature of the universe is incomprehensible, so much so that it's less the domain of science and more the domain of philosophy. First of all, the universe is vast; so vast that just thinking about it can give you a headache. To get to the edge of the solar system (the tiny place that houses the planet we call home) it will take one year, and that is only if we travel at the fastest speed physics-ly attainable, the speed of light. How fast is that? Well, a ray of light can do thirty round trips between Boston and San Francisco every second. So it would take you a full year traveling at bizarre speeds to get to the edge of the solar system, which itself is a minute piece of the universe. How minute? Let's see: there are billions of stars, just like our solar system, in the Milky Way galaxy, and then there are billions of galaxies in the universe. If this gigantic Solar system was one speck of dust, the universe would be as big as all of Earth.

So it should come as no surprise that there are so many unanswerable questions about the universe, and even when answers start arriving, I am doubtful that we'll be able to grasp them. For example, what are the boundaries of the Universe? If you stand at the edge of the universe and look outside, what do you see? In fact, if you stretch your hand outside that edge, does it disappear, or does that suddenly become part of the universe and the edge moves further out? How far does that nothingness stretch and where does it end? I suspect these are the same types of questions that philosophers and thinkers had about Earth before they realized that it was round. That roundness answered the end-of-Earth question satisfactorily and diverted the average man off thinking about the extent of our universe, but the fact remains that the question has not been answered, and I suspect will remain open for a long time. Another question that goes in the opposite direction is: How small can things get? We now know that atoms exist and inside them are the electrons, protons and neutrons. Furthermore, we know that protons and neutrons are made of quarks and potentially quarks might divide further into strings or whatever subcomponents make fundamental sense. The question remains: Where do we stop? No matter how small we get, there will always be smaller, just like the universe problem: No matter how far we go, there will always be further.

Even more perplexing is time. How long will the clock keep ticking? Will there be an end? How can there be an end when anything that happens *after* the end is clocked by time? Going back in the opposite direction, when did time start ticking? Was it after the big bang? What about *before* that? Was there nothingness, and if there was, wasn't it clocked by time? It's all puzzling and all attempts to explain it have been pure philosophy. I remember one time I picked up a book by Stephen Hawking called *A Brief History of Time*[2]. I bought the book because I wanted to learn what one of the greatest minds alive thinks about how time started. After the first chapter, I was thwarted. Hawking explained that the most impacting incident in the life of universe that we swim in was the Big Bang. Anything before that is un-speculate-able and irrelevant to the scope of the book and therefore will not be discussed. I would think that Hawking avoided the subject altogether because he could not even find a single clue, not even an infinitesimally small one, that could give hints and help in this guessing game. And guessing is probably the most fitting word when it comes to describe the science of the cosmos. All the wild theories about origin, size and future of the universe are based on our ability to

---

[2] Stephen Hawking (1988). "A Brief History of Time". Bantam Books. Bantam. NY.

see it. And the sources of that ability lie in the electromagnetic waves (visible or not) that shower Earth constantly. To put it in simpler terms, our understanding of how the universe works comes from looking at its objects from a distance – a fantastically large distance. It's like asking a passenger on a thirty thousand foot high flight to guess the number of leaves on the trees below when passing over the Amazon. Rest assured that whatever theories were formulated and even accepted about the nature of the universe, will be tested ungracefully in the future.

A Science Dialogue, With a Twist

There is no ambiguity in the message that I am trying to convey which is that science is not perfect. We consent to extensions of solidly firmed concepts and accept irrational, anti-commonsense paradigms in order to make good use of the models that produce good results. And that is not necessarily a bad thing. In fact, that is exactly what we need science for. From Einstein's theories came the $E=mc^2$ formula, and eventually nuclear power and more. Will there be a better model that explains why Michelson and Morley could not find their fringes? Will that model explain all the consequences of Relativity without introducing weird concepts like time-space continuum? Potentially, yes, but for now, we'll accept what Einstein dished out and use it most advantageously. And Science will continue to be that useful tool that will help us decipher our universe.

So what fuels this science charge and where is it heading? Ray Kurzweil, a very credible futurist, has successfully and over multiple iterations predicted the technology trends of the future for the past two decades. He foretells a future that transforms beyond the imagination of the wildest science fiction writers. A future that starts with the merger of man and machine and ends with an overwhelming intellectual power that spans the space around us and grows like fire in a forest, eating energy in the universe – energy in matter or any other form—and expanding outwardly at speeds that we can't even fathom. It all reads like a science fiction book, and not an even a mild one, but believe it or not, this could be the path of the world.

Will all of this happen and will it be detrimental or beneficial to humanity? This will be the subject of this book. The rest of the chapters are mostly dialogues about this theme. Each chapter starts with an introduction to the subject at hand and tails into a dialogue. Four main topics are discussed:

- Science Eccentricities: Why black holes suck, Relativity is about absoluteness, and the **uni**verse is **multi**dimensional. In general, the discussions will be around why Feynman in his QED book explicitly stated the convoluted nature of physics[3] *"This is the third of four lectures on a rather difficult subject – the theory of quantum electrodynamics -- and since there are obviously more people here tonight than there were before, some of you haven't heard the other two lectures and will find this lecture almost incomprehensible. Those of you who have heard the other two lectures will also find this lecture incomprehensible, but you know that that's all right: as I explained in the first lecture, the way we have to describe Nature is generally incomprehensible to us."*

- Robotics: How some want machine to take over, others want machines to go away, yet everyone is in an urgent

---

[3] Richard Feynman (1985) "QED, The Strange Theory of Light and Matter". Princeton University Press

Introduction

rush to surround himself with as much electronics as possible. The debate heats up to a boiling point on the topic of machine intelligence and its effect on humanity.

- Bio-Universe: How our age old pursuit to understand the intricacies of the inner space will be trivial compared to what is about to materialize in the coming few years, especially on the aging-combat front. Although the aging topic has been tackled profusely over the centuries, with no encouraging results, Aubrey De Grey's *Ending Aging*[4] book legitimizes the discussion and moves the whole topic from voodoo and black art domains into the science domain.

- The Future: Why a trip to the future might be much more complicated than Michael J. Fox's time machine journeys. The discussion heats up on paths of the future; the relatively near future. This is not a Star Trek captain's log year 2430 discussion. This is about the radical transformations that will come about in our lifetime, potentially as early as 2030. Although everyone agrees that the future will be bright, some are worried about a different, more intense, meaning of the word bright.

The cast includes Herbert, a brilliant physicist, Buzweil, an optimistic futurist, John a realistic conservative, and me.

Herbert eats, drinks, and sleeps science. He is competent and knowledgeable but has deficiencies on the sense of humor front, and to make things worse, he is completely unaware of his deficiencies. He has great passion for science and

---

[4] Aubrey de Grey (2005) "Ending Aging: The Rejuvenation Breakthroughs That Could Reverse Human Aging in Our Lifetime". St. Martin's Press. NY.

unbounded respect to scientists, so much so that an attack on a science icon is nothing short of blasphemy to him.

Buzweil is an inventor, a technologist, and a rogue advocate of the idea that the future of applied sciences is the future of humanity. He is impatiently waiting for the technology transformation engines to mature so he can ride the future wave. He is also a firm believer that the path of the future is governed by technology and he is optimistic on the direction of that path.

John is a conservative executive, reserved in his opinion on science, traditional with an unambiguous technology-skeptic attitude, and a strong opponent to radical changes that could have drastic impact on our existing way of life. He is intelligent, very well rounded, and can make and defend an argument better than a lawyer.

Finally, and for no apparent reason but to write the book, there I am in the middle of every discussion, mostly to mediate and keep the discussions civil, and occasionally to act as a punching bag for those in the room who find themselves in need to vent some steam. To describe myself, I...

...well, you'll find out.

To say the least, it's an interesting collection of amigos, with widely varying backgrounds tackling some very tough topics. Put them all in one room and you will unmistakably get educated, entertained, and probably black-eyed.

2

SCIENCE ECCENTRICITIES

Science is the outcome of the relentless pursuit by a fanatic minority to reveal what's hidden in the deep cavities of nature; the elusive, shy and sometimes singular nature that refuses to unfold and show its true colors. Science is appealing to some, appalling to others, and was ignored by most until it started to transform every facet of our civilization. Nowadays, you cannot imagine surviving without the science-shaped necessities of life; necessities that were luxuries a decade ago, dreams a few decades before that, and wild imagination last century. In the nineteenth century, if you were a visionary that managed to concoct a legend about rockets to the moon, black holes in space, and aircraft carriers, you would have made a great comedy act. Today anything goes. You ask the average man about the plausibility of a Star Trek-like future and you will be reprimanded for asking. The theme of the day is: '*If*' is dead, long live '*When*'.

So needless to say, it has become an undisputable fact that science is a driving transformation force in our society and a banner of exactness in our lives. Nonetheless, questions linger about the correctness of the science we have been putting on the table recently. Altercations and controversies have become the norm in scientific circles, fueled by some theories that require quite a stretch on the imagination. It's as if scientists have reached a mental block and are trying anything to break the monotonous lack of progress. It is a fact that, for the past fifty years, the physics world has experienced severe deceleration in finding breakthroughs at the fundamental constitution level. We discovered the electron in the nineteenth century (1897 to be exact) and it took close to seventy years to find the next set of granular particles, like quarks, leptons and bosons. Fast forward fifty years, and you'll find scientists building $10 billion particle accelerators to discover the next level of details. As we peel more onion layers, the chase is getting harder and the questions are getting tougher. For example the age-old quest for time and space boundary conditions has hit a brick wall.

> *Time*
> The two directions, history and future are still complete unknowns. We are unable to build any theories on the history of time, even when the most brilliant man alive wrote a book about the subject. And future predictions are not even in the realm of

science, except for some rudimentary applications like weather forecast.

Space
The two directions, outwards and inwards, are also as murky. We could not make a dent on building any credible formulation that describes how our macro-universe is bounded. Outer space is still the most mysterious part of our being. And just as hard is the search for answers at the micro-level. Particle physics, which is supposed to enlighten us on how the fundamentally small particles behave, is not only perplexing and mysterious, but is also utterly incomprehensible.

Even Einstein was stymied. He spent the better part of his life looking for a unified field theory and came up empty handed. His crusade continued by an army of the finest minds of the twentieth century, but the theories that they created were, by all accounts, incomplete and controversial.

No one is throwing the towel though. The pursuit for science exploration is more zealous than ever and investments in science are growing at a fast pace. Aided by some marvelous tools of productivity, the science drive is thriving more than ever. The only worrisome thought in this whole matter is what we will find next. When we *cracked* the atom, we were able to craft monstrous tools of demolition with unimaginable destruction capabilities. Some part of me wishes that the *cracking* stops at the atom level. It just might be safer.

The Dialogue

Monday 7:00AM. The sun was shining, the birds were chirping, and everyone in the world was greeting the birth of a new day. Why they had to do it this early in the morning, I have no idea. I am the 10:00 AM type. Ten is the perfect number that begs to be coca-doodled to, and chirped at; the number that truly deserves to be called the beginning of the day. I tried to convince my boss of my ten-to-five working-hours theory with the embedded messages: Ten is one and zero, the binary digits that run the information technology world. Ten divided by five is two clearly implying twice the productivity; but he couldn't understand the cosmological implications.

I woke up and rushed to the shower in an effort to beat the PM conundrum (see chapter two). Then autopilot took over the system: *shower, shave, after shave, deodorant, tiptoe to the closet to avoid waking the baby, put clothes on and head to the kitchen for the cup of coffee*. I finished the first caffeine dose of the day, grabbed the laptop and barged out of the house and into a fresh week.

In the office, and true to song, Monday started to manic right away. As soon as I stepped in, I was hostile-greeted by the other strong woman in my life, my assistant. With one hand on her waist --a self-improvement technique that she apparently learned from a course that she must have taken with my wife— she looked me straight in the eye, and very scornfully said.

"How many times did I tell you not to mess with the calendar? You are double booked, *again*, at ten and there isn't even a name or contact number for the surplus meeting."

And trying to ease things out I replied.

"Am I? Ok, we'll sort it out, but first things first. Good morning; how was your weekend?"

"Don't change the subject. What is this BS meeting that you added over the weekend?"

And I resented that. There is a line that an assistant cannot cross.

"Listen, my meetings are not BS. I must have done the double booking for a reason."

"Really? Well, this meeting is BS. You scribbled the word BS on the title yourself, look."

She pointed to the screen and I looked at the calendar to see what she was talking about, and sure enough the letters B.S. were in the timeslot. After a little while of digging into the memory bank, trying to find out why I put that meeting there, I remembered all.

"Ah, yes, it came back to me now."

"*Came back*? How on earth did it ever have the chance to *leave* in the first place? You added the meeting last night at nine PM. Have you installed a revolving door between your ears or is it built in?"

That I could not defend. I fiddled with my car keys for a second and then said.

"Well it was a long weekend and I had so much to do. Anyway, it's not BS. It's Broadband Systems. It's just a follow up call that I have to do with the team."

"It sounds like a BS meeting to me."

She then pulled a stack of papers off her desk and gave it to me finishing her sentence.

"You know, I have no idea how they made you boss; now run along to conference room 4C, your 8:00 AM is waiting."

I know what you must be thinking. Who is the boss in this setting and for what convoluted reason do I keep an assistant like that? Well, she might be rough on the edges, but she is extraordinary. Before her I was all over the place. Meetings were forgotten, flights were missed, and occasionally, full day calendars were dropped. Now I can sit down and relax knowing that I am in safe hands. She is that good. I just have to take the good with the bad.

I headed to the fourth floor and rushed into the C conference room. Michael and Susan were there, apparently running a before-the-meeting meeting. For those of you who don't know Michael, don't. Seriously though, Michael is one of those I-am-well-weekend-rested folks. He is always in the meeting ten minutes before the scheduled time, with a cup of coffee in one hand, a notepad in the other, and a view-only laptop in the middle. I good morning-ed them both, did the usual how-was-your-weekend chat and sat down waiting for the rest. Once we had quorum (the most beloved word in the office environment) Michael, who by now had finished his second cup of coffee and his caffeine jolt was bottled down ready for uncorking, uncorked.

"Folks, before we jump into the discussion at hand, we had an issue with the installation of GEORGE release 1.2.2.1.4.4.5e that we need to go over."

GEORGE is the latest, cutting-edge, state-of-the-art, object-oriented, multi-threaded, fault-tolerant, backward compatible, forward extendable, ISO-9000, ordering system. It stands for Generalized and Enhanced Ordering and Reporting Gateway Engine. In IT, why we are obsessed with making sure that acronyms look like proper words, I have no idea. Like when you hear the word GEORGE you immediately think *"Aha, right, I got it, that's an ordering system"*. Of course we don't build systems or programs; we build the much more sophisticated Engines!

Anyway, that's how Michael started the meeting and when Susan replied, the dialogue that followed was enough to warrant legalization of the use of well stuffed, gently rolled, marijuana joints.

"I thought we cancelled 1.2.2.1.4.4.5e."

"No we cancelled 1.2.2.1.4.4.5a and merged 1.2.2.1.4.4.5b and 1.2.2.1.4.4.5d."

"What about issues 44, 46, 49 and 52 on 1.2.2.1.4.4.5b."

"Issue 44 was closed by SIT, verified by UAT, and tracked in the books by PMO. Issues 46, 49 and 52 were transferred to 1.2.2.1.4.4.5d as issues 17, 19 and 23. "

"And what happens to 1.2.2.1.4.4.5c."

"1.2.2.1.4.4.5c will be combined with 1.2.2.1.4.4.5f. That should simplify our release process."

And then Michael spent the next two hours explaining how the release process with the new change was *simplified*!

That's how my week started. I practically spent the day jumping from one conference room to another, like a student running between lecture halls, attending meetings, workshops, and sessions and expanding my office acronym vocabulary. Finally, I wrapped up the day with a meeting with my assistant, who seemed in a better disposition than she was in the morning. Apparently, seeing me drained had a soothing effect on her mood. She fed me all the calls and messages and we rearranged the calendar to fit a couple of urgent meeting-requests from the big guy. This always happens on Monday. As you know, in order to be competitive in the fluid market environment, you'll have to be agile and fast-responding to industry changes, realigning your strategy to ever improve the competitive stance of the enterprise. At least, that's the official statement. In reality, however, the boss, who is running with an inconsequentially marginal strategy made mostly of industry buzzwords, wakes up Monday morning dreaming up new amendments that need to be applied immediately.

At around 4:30 PM, I was getting restless, impatient and PMconundrum-ed. I started packing my stuff and was ready to bolt, when I got a call from John who invited me to join him, Buzweil and Herbert to an end-of-the-day cigar in Cool River next door.

Science Eccentricities

I called my wife to tell her the bad news, but was surprised to receive a very understanding *'sure honey, have fun'* response from her. I packed my laptop bag and was on the road in two minutes. First on the scene, I headed to the infamous cigar room and got a head-start.

Buzweil and John arrived within five minutes and Herbert soon afterwards, and as soon as I saw Herbert I knew he had some announcement up his sleeve. He did not fail me.

"People, I have some news. I got notification today that my paper on abstract micro-representation of the human genome was accepted for publication."

"Good for you Herb; this is amazing." I replied, and then continued "what is your cro-representation and how is it abstract? Is it because it's not really an exact depiction of the form?"

John, who had not lit his cigar yet, looked at me pompously, with the cigar in his mouth and the lighter in his hand ready for ignition, and said.

"It's micro-representation not my cro-presentation you IT imbecile, and it's abstract as in pure and mathematical not abstract as in a Picasso painting. You IT geeks, if it does not have RAM, Byte and Giga Hertz, you have no idea what it means."

I shrugged my shoulders and said "Whatever" and he turned to Herbert and continued.

"Everyone is genome-ing these days. DNA has become the central theme to so many sciences. Even nano-technologists are riding the bandwagon and building hopes on creating self-assembling machines using a derivative of the DNA."

"Well the field has great promise. It turns out we can learn so much from the way our biological system assembles itself into trillions of self-governing compartments that work flawlessly. And it all starts with one cell."

"If you ask me, we are starting to latch onto the most trivial of hopes because we are closely approaching our limitations. Science has run out of gas. We pretty much know everything we will ever know. The rest is not within reach."

Buzweil, who by now had pushed a couple of puffs through the respiratory pathways, was ready to engage.

"On the contrary; I think we are just about to launch the most fantastic scientific voyage since the industrial revolution."

John rolled his eyes before he turned to Buzz and said.

"Really? Have you seen what we have been dabbling with in the twentieth century? Some of the theories that scientists created in the past hundred years are fabricated nonsense."

"Like what?"

"Like Relativity."

"What's wrong with Relativity?"

And John, as if bored with the calm status and eager to ignite the discussion, along with his cigar, flicked the lighter, lit his smoke-staff and went deep; probably too deep; definitely too deep for me.

"The easier question to answer is probably what is right with Relativity. Relativity is a theory that, not only dishonorably discharges fundamentally accepted axioms that so many think-tanks had come to establish over hundreds of years of intellectual diligence, it also takes aim at our basic common sense. Relativity is the first theory that accepted the unacceptable. In mathematics, you remove the impossible scenarios, the contradictory ones, the ones that defy the basic axioms, and you are left with the only possible outcome."

And I saw the analogy, at least my kind of analogy.

"Sort of like Sherlock Holmes. *When you have eliminated the impossible, whatever remains, however improbable, must be the truth*?"

"Yes Sherlock, sort of like Sherlock Holmes. Relativity, however, breaks the basic fundamental axioms with no regard to truth. Driven mostly by our inability to find satisfactory solutions to challenging problems, we bring Relativity to life, followed closely by some of the weirdest science ever to make it into science books."

After hearing this, Herbert, who had faded into the background for the last couple of minutes, went into a state that can only be described as a combination of startled, offended, and irate. With a serious tone he came back into the conversation.

"These are very harsh words about one of the most popular theories in science; a theory that ushered a new age for physics and breathed innovation into the world."

And I could not agree more.

"I know. John, what do you have against Relativity anyway? Were you like orphaned by an Einstein book falling on your mother's head?"

Herbert shook his head at me as if saying *would you cut it out* and continued.

"Without the beautiful models that were built by Einstein, Bohr, Schrodinger and the rest of modern physics pioneers, we would not be celebrating the intellectual festivals that we do in every science book."

Although I agreed with every word that Herbert said, I still had some reservations.

"I tell you what though John, I agree with you on one thing. Physics 212 would have been a much easier course if it were not for these convoluted concepts that I could never understand."

Apparently, that was the line John was waiting for to dive where I really did not want him to dive.

"You are not alone my friend. You will probably never meet anyone who fully understands what Einstein did to science. According to Relativity, objects become heavier as their speed increases. To be exact, their mass increase by a factor of $\frac{1}{\sqrt{1-\frac{v^2}{c^2}}}$."

Science Eccentricities

And my left eye started to twitch uncontrollably as soon as he said square root.

"College flashback, and not a good one either. If I really wanted to walk down the memory lane, the last place I'd like to go is professor Gainer's physics class. Can we do without the formulas please? I had enough of those to satisfy my intellectual curiosity for a lifetime."

Trying to *simplify* things, John changed the mode of the discussion from hard, to utterly, resentfully, complex.

"It's really not rocket science; it's just weird. C is the speed of light and V is your speed. As V gets closer to C, V/C becomes almost 1, and $1-V^2/C^2$ becomes almost zero. 1 divided by a number that is almost zero is a big gigantic number. That's how much your mass will grow when you dash at the speed of light."

"That's why I hated physics 212. Every time you look at one of those formulas, you'll have to flip your brain a few 180 degrees. V increases, V/C increases (flip number one, tilt head), V^2/C^2 increases (flip number two, tilt to the other side). $1-V^2/C^2$ decreases (flip number three, tilt again). $1/(1-V^2/C^2)$ increases (flip number four, another tilt). By the time you finish reading the formula, you feel that your brain has done three and half back somersaults."

Herbert was shaking his head while I was swaying mine.

"Don't be silly, it's straight forward. V goes up, mass goes up. When you travel at half the speed of light, your weight will double. When you travel at 0.99 the speed of light, your weight will increase by a factor of ten. If you travel at the speed of light, your mass will become infinite. That's one reason why nothing can reach the speed of light. You will need an infinite force to move you one inch if your mass is infinite, and infinite forces do not exist."

"Herbert, my problem is not with the mathematics of the model. My problem is with the consequences. After Relativity, it became acceptable that mass increases with speed, that nothing can go beyond the speed of light, and, eerily I might add, that travelers age slower than static people. With this foray of paradoxical concepts, Einstein opened an uncanny door for scientists to wander where no self-respecting intellectual had ever treaded before. Heisenberg introduced his Uncertainty Principle which states that there is a limit to measuring speed and position of an object. The more accurate your measurement is for one, the less accurate it is for the other. So if you can measure, precisely, the speed of an electron, its position could be anywhere in the universe. Remember, this is not a statement on the capabilities of state of the art measuring devices. This is a physics theory establishing boundaries of nature; boundaries that we will never cross."

Now that was strange and I couldn't understand it at all.

"But wait, what does that mean? I know your position and your speed right now. You are sitting on the couch and your speed is zero. You are not moving; does this mean you could be on the moon at any instant? That is without Honeymooner's Ralph zooming you there of course."

Herbert, like a mother explaining why one and one is two, put his hand on my shoulder and said.

Science Eccentricities

"My friend, you are moving. Every particle in your body is moving constantly. Electrons are in constant flux around the nucleus. Atoms are incessantly moving, or at least vibrating inside solids. That's why you are not on the moon."

Then he turned to John and continued.

"Don't put Heisenberg's Principle down. It's one of the most important achievements of the twentieth century. It helped plot a course to create the most important field in science, quantum mechanics."

"Herbert, I am not putting it down. I know that it helped scientists model the world and formulate structures that concluded with amazing discoveries. All I am saying is that it does not make sense."

"It does to me. There is a rationale as to why we can't measure with exact accuracy; we just don't know what that rationale is. You see that's the beauty of science. It's like football. You shoot the ball few yards down the field and someone picks it up and runs with it; and then he shoots it a few more yards, and so on."

"I would say the ball was shot clear out of bounds. Check out the Big Bang theory for example. Everything in the universe came out of one tiny infinitely small, infinitely dense energy entity. That energy monster, that very tiny energy monster, exploded to create the universe with everything in it. Our entire universe came out of a speck. Go tell that to the average man and see what response you'll get. To say the least, it's utterly absurd."

"The Big Bang theory came as a culmination of the works of so many scientists that strived to understand the beginning of everything. You quickly skipped through so many iterations that led to the creation of this beautiful theory."

"Herbert, be accurate in what you say. The Big Bang theory tries to explain what happened since the Big Bang. How everything started is not even a science problem yet. Stephen Hawking in his book *History of Time*[1] clearly states, and to my disappointment I might add, that the scope of his book stops at the Big Bang and since there are no clues as to what happened before that, it will be considered irrelevant and outside the realm of his book. In fact, I am sure it's outside the realm of science."

Herbert adjusted his seat and got on the defensive. Whenever someone takes a shot at science, it becomes personal to him.

"There is nothing outside the realm of science. Science is the universal tool that unravels the nature of all things. We simply have not attained the skill and knowledge to unlock the origin-of-the-universe mystery."

"Typical. Hide behind the walls of in-progress. Just do me a favor and accept that the boundary problems are going to stay intractable. We will never know what was before, and we will never know what will be. Also we will never know what's in the microscopic world and we will never know what's at the edge of the universe. We are simply not equipped to know everything."

---

[1] Stephen Hawking (1988). "A Brief History of Time". Bantam Books. NY.

Science Eccentricities

And to my surprise, Herbert, in agreement with John, said.

"Precisely, and that is why some of these formulas and concepts are hard to understand. Once we prod into places outside the norm that our *senses* are used to, things might not make *sense*."

There was something to that and John was clearly taken aback by the last statement. He was expecting a defying comeback and he got a lame concurrence. However, that did not deter him.

"You can't just accept absurd concepts in the name of sense deficiencies. These theories might be models that work momentarily, but I know that someone else will come in the future with substitutions that work just like the old ones, without the preposterous side effects. Let me give an analogy. Galileo, Newton and the whole scientific community before Einstein would agree with you(if they were here) when you say that an object traveling at fifty would seem to travel at twenty to someone moving at thirty , correct?"

And even I could follow that logic. I leaned forward, as if I was running the show, and said.

"Right, I got that. What's your point?"

"Einstein came in 1905 with his Relativity formulas and showed that the correct observed speed of the moving party is actually greater than twenty. The difference was too small to notice, or even measure, but nonetheless it was there, right?"

Now that was outside my league, but apparently not so for Herbert. I leaned backwards and let Herbert take the lead.

"Alright, so I'll ask the same question, what's your point?"

"My point is that formulas that seem unequivocally correct can be overturned and superseded by more accurate ones. New formulas will retain all the properties of the old ones, but will add more to explain new concepts."

Backward compatibility; the new version of software works with the old dataset but adds more functionality. I was happy to put my two cents on the table.

"It's like version 2.0 of the formula, works with the old, and covers new grounds."

"Again, if it is not in IT jargon, you have no idea what it means, do you? Yes, just like backward compatibility. Now I expect that in the future we will produce such theories that can explain everything that Relativity did, but will cleanse science from the senselessness and incomprehensibility that mired the physics world."

And Herbert, amiably replied.

"I will not disagree with you on this hypothesis."

With that declaration, Herbert seemed to have found a rhythm with John. A truce seemed imminent and peace appeared close at hand. Herbert sat back, grabbed the lighter and ignited his ignored cigar; then he dropped a grenade on the table.

"In fact, a most befitting example to support your argument is String Theory."

Science Eccentricities

Looking at John's face I realized that the brief truce that lasted for a few seconds was over. His eyebrows came together and his demeanor turned unpleasant before he lashed out on Herbert.

"Don't get me started on String theory; the Theory of Everything. What was supposed to be the apex of the intellectual powers of humanity turned out to be the most irrational concept I have ever come across."

"What is String Theory?" I inquired. I was totally unaware of the bugger.

And as if I was on mute, Herbert turned to John and said.

"My god, why do you have so much hatred for what brilliant people do to advance science?"

I didn't like being ignored that way, so I leaned further forward and asked one more time.

"What is String Theory?"

Again, I got the deaf ear and felt like a child nagging "mom, mom, mom" when his mother was totally submerged in a juicy gossip piece with the neighbor.

John, apparently agitated by what Herbert said, raised his voice a notch and said.

"I don't have problems with brilliant people doing brilliant work. I have a problem with science circles accepting, as facts, models that are ridiculously absurd."

Shifting Borderlines

And that was it. I stood up and, almost screaming, asked yet one more time.

"What is this crazy String Theory that you cannot stand."

John finally noticed me. He turned around and started to vent right away.

"String Theory is supposed to be the zenith of physic. According to Brian Greene[2], String Theory was Einstein's dream of unifying nature's four types of force fields into one grand force."

"One grand force? As in the force that Yoda keeps referring to when he says *may the force be with you*?"

And I wasn't expecting much return on such flimsy joke, but apparently I did not know Herbert that well. That bit cracked him up. He kept laughing for two continuous minutes and I was glad to see him feel better.

When he finally was able to catch his breath, he returned to his serious tone and said.

"John, String theory, if validated, will actually get to the lowest granularity of the universe. It's not just about unifying the fields; it's about unifying the particle world and the basic constituents of the universe."

I still did not know what this String Theory was about and the discussion was too heated for me to ask again, so I grabbed the open laptop and looked up String Theory. There were tons of

---

[2] Brian Greene (2003). "The Elegant Universe". Vintage. NY.

hits, but at the end, I found a nicely written lecture by Brian Greene[3], a theoretical physicist from Columbia, that expressed the concept nicely.

> *"The fundamental particles of the universe that physicists have identified—electrons, neutrinos, quarks, and so on—are the "letters" of all matter. Just like their linguistic counterparts, they appear to have no further internal substructure. String theory proclaims otherwise. According to string theory, if we could examine these particles with even greater precision—a precision many orders of magnitude beyond our present technological capacity—we would find that each is not pointlike but instead consists of a tiny, one-dimensional loop."*

If that actually turns out to be true, it's a cause for celebration. It's a graceful way to describe the lowest common denominator of nature. I couldn't see why John had so much bottled dismay for something that could actually be nicely comprehensible. And I wanted to know.

"John, baby, I see nothing but splendor in this Rope Theory. Why is it on your black list?"

"It's String Theory you dim-wit. It is hanging by a *rope*, if you ask me though. The reason it's on my black list is because it predicts that we live in an eleven dimensional universe that fits on one of infinite number of branes, whatever that means. To take this

---

[3] Brian Greene. "Challenging What We Know - Superstring Theory". *Augustana College, Rock Island, Illinois*, May, 2004

one step further into idiosyncrasy, our universe was created when two such universes came in contact. A little hug, a kiss, or an embrace, whatever you want to call it, and voila, one universe coming up ready for serving."

Now that was weird and I saw why John was skeptic about the whole meal.

"I don't understand. The first part is nice. Combine fields, combine particles and find the foundation of physics. Why do you have to go supernatural?"

Herbert sighed before he answered that question. He looked like someone on a witness stand, under oath, testifying how he saw his best friend steal exhibit number one.

"Because that's what the mathematics dictates. We didn't ask for or pursue eleven dimensions. They came out of the complex formulas. It was either that or particles having negative probability for existence, and negative probabilities do not mean a thing."

And I just could not understand this guy. Eleven universes is the alternative? If I were the leading String Theory physicist, I would chuck both options into the trash-can and start over. I would probably drop the field all altogether and go farm some piece of land on the outskirts of Albany. But that was just me. I am sure that these brilliant physicists had good reasons to continue with their hunt.

John, moving from one lead to another, continued his relentless pursuit like a fine Lexus. You see, once John gets on a track, he goes all the way. He was enjoying all of this and he wanted to squeeze as much juice as he can.

"And this is just one of many concepts that littered twentieth century physics. Take for example the graviton proposition."

I had no clue what a graviton was, but I knew immediately where I heard the term. It sounded like something Jean Luke Picard would say to Jordi after the hostile ship uncloaks on the bow and right when the shields take another hit and go down to 7%: *Jordi lower the shields and fire the Multiphase subspace graviton beam at the Klingon vessel.* Of course, no respectable Start Trek episode is complete without the weakening shields getting bombarded multiple times by the enemy vessel. Still, I had no idea what a graviton is.

"What's a graviton?"

With his hands up in the air, clearly showing his absolute despise for the subject, John answered.

"It's a theoretical particle, i.e. a particle we think exists! It does not have any mass or charge and it is conjectured to be responsible for gravitational forces."

That was the kind of tone that Herbert had to deal with all night. I, myself, was indifferent to John's attitude, but Herbert and physics are lovebirds. He raised his voice and with a touch of spite asked.

"And what is wrong with that?"

That's about as nasty as Herbert can go.

"Herbert, we can't see this graviton, we have no hope of observing its effects and we don't even have a thought experiment on the subject. Yet, we'll have to accept it."

"Well, the same applies to photons. They are particles that have no mass and are invisible individually, yet we came to accept them."

"No, photons are not the same. Einstein won his noble prize on the photoelectric effect that experimentally linked the photon to quanta. In the graviton case, we don't have any such link to reality. Graviton is a hypothetical elementary particle that transmits the force of gravity and it will stay hypothetical until we run an experiment that proves its existence. Even theoretically, we have enormous hurdles to overcome before we can treat this graviton as a proper particle. Although we think it's responsible for the gravitational force, we have no idea how it does that. What type of exchange of graviton happens between objects to create the gravitational force? These are very tough questions and it would require a great deal of theoretical efforts before we can build a model. Detecting a graviton is another big problem. It has no mass and an extremely small amount of energy."

From the look on Herbert's face, I could tell that John had made some good points. Herbert leaned back for a few seconds, scratching his head, as if he really did not want to continue with the discussion. When he talked, all he could come back with was a wish list.

"John, we will eventually get the technology that will be able to dig deep into the micro world and detect the presence of gravitons and whatever sub-particles make up the basic constituents of matter."

Of course that did not stop John.

"It is going to be a very long time before we have something that can detect gravitons. The Large Hadron Collider (LHC), the world's most powerful particle accelerator that was constructed in Europe and was active in September 2008 took years to build at a cost of $10 billion dollars. You will need tools a few orders of magnitude more powerful than that to start contemplating an experiment that can detect gravitons."

At that moment, Herbert's phone rang this weird ring that had Herbert written all over it. He answered the call and walked out, and I picked up the discussion from there.

"What's a Hydrogen Collider?" I naively asked.

"Its Hadron not Hydrogen. It's a particle accelerator. It sends elementary particles racing into a tunnel and smashes them against each other at speeds close to the speed of light. Such head-on collisions can break the particles into sub-components, if they exist. It's our only weapon to discover how small things can get. It was built for physicist to prod further into the atomic world in search for elementary particles. "

"Got it. You try to whack particles with a hammer so you can blow them up to check what's inside, but you are stymied because you don't have hammers that can crush teeny thingys. Instead, you thrash them. You haul them around and ram them against each other, right?"

"Whoa, far out; awesome. Dude you totally rock. I like no way could have said it funkier! You're the man."

Then he paused for a second to get off the sarcasm channel.

"I have to tell you that I love your broad command of scientific terminology. But yes, that's the general idea *dude*."

I totally ignored the derisive implications. I just wanted to know what it is we were still looking for. I was under the impression that physics had established the basic ingredients of atoms. What is still to be found down there?

"I thought we knew what particles are elementary. We even teach that in elementary school now. Protons and neutrons are crammed in the nucleus while electrons fly freely outside."

"Well, electrons might be elementary particles, but protons and neutrons are not."

"What do you mean protons and neutrons are not? That's all what we studied throughout high school. In the atom, negatively charged electrons balance the positively charged protons, and then roughly the same number of uncharged neutrons nestles next to the protons. If these are not particles, then what is?"

"Elementary particles are the tiniest building blocks. A particle that is made of particles is not elementary."

"You mean to say a proton is made of smaller particles?"

"Yes. In the Standard Particle Model all mass is made of twelve particles. Six of them are quarks and another six are leptons."

"Leptons and quarks, sounds like an ad for a religious tea brand. Would you care to elaborate and unravel for me the elementary contents of the micro-world?"

And I looked at John to see if he noticed that my vocab was not all trash. He smiled.

"Leptons are the easy ones. There are six types of leptons, all variations of the electron. Think of leptons as the electron and five of its cousins. In fact, it's more like the electron and two cousins (all negatively charged) and then three neutral counterparts. So, you have the negatively charged electron and its neutral counterpart, the electron neutrino; the first cousin, the negatively charged muon and its neutral counterpart the muon neutrino; and finally, the negatively charged tau and its neutral counterpart the tau neutrino."

And quickly, my superior deduction skills went to work. I knew exactly what was coming next.

"Let me guess, quarks are protons and its cousins."

John smiled briefly. I could almost hear him thinking, *he fell for it*. And with the satisfaction of a hunter unloading his trapping net, he answered.

"Good try, but no. There are six types of quarks, up and down, top and bottom, charmed and strange. For example, proton is made of two up and one down quarks. Neutron is made of two down and one up quarks."

That was borderline R-rated and exceedingly convoluted. I did not like it a bit.

"So you have twelve elementary particles that make up the three particles we know!"

"No, you have twelve elementary particles that make up the hundreds of particles we know."

"Hundreds? Since when?"

"They have been coming along since the early sixties. I am sure more will be found, and that is why we needed a more fundamental model."

That was a definite put-off. Models were supposed to simplify things. The old stuff we took in high school made more sense. Three types of particles, two in the nucleus and one outside, were more than enough for me. My heart goes out to the high school students of today.

John, like a missionary preaching the ways of God, continued with his sermon.

"You see, we keep finding new ways to change the world. Demolition tools keep blasting away old models and constructing new ones. However, the new models are spookily haunted with strange concepts that do not make sense. Science is plateau-ing into strange zones."

I still thought that John was blowing this out of proportion. So what if we had few theories that were out there. Brilliant people tend to be a bit eccentric.

"You can't generalize this to all of science. I don't think there is a trend here."

"Yes there is. Unconventional theories are popping up like popcorn all over the place ever since we started to hear the terms Relativity, Uncertainty and double slit experiments."

I guess science is full of surprises. That last bit sounded like an Alfred Hitchcock movie and I had to ask, but John's reply was disappointing.

"The double slit experiment is a perplexing phenomenon that probably caused sleepless nights for many, and the explanation by the physics community is, to say the least, an enigma on its own. The setup is simple. A gun that fires electrons on a screen with two slits. Behind the screen is another screen that collects the fired electrons. Given that we are firing *particles*, and not *waves*, the collected electrons on the back should not create interference fringes, but they do."

"What's a fringe?"

"Interference fringes are bright and dark stripes that occur when you run an experiment like this with light instead of electrons. You see, light, acting as a wave, goes through both slits and sends two different waves towards the screen behind. Wherever the top of the wave coming from slit one meets a top of the wave coming from slit two, you get a bright spot, and of course when a top meets a bottom, you get a dark spot. With the two waves arriving at different levels on different locations on the screen, you end up with dark and bright stripes, otherwise known as interference fringes. The experiment is easy to reproduce, and shows clearly that light acts like waves.

But for this to happen with electrons is, to say the least, a puzzling phenomenon. And don't hold your breath for a satisfactory explanation; it is more perplexing than the phenomenon itself. Apparently matter, including light and electrons have both wave and particle properties. Sometimes it acts like a wave, other times like particles. "

Herbert had just stepped back in when John started his hoedown on the particle and wave properties of matter. It was clear that he did not like what he was hearing. In fact, from the look on his face, and the tone of his voice, you could almost feel that he wouldn't mind running a double slit experiment on John, if you know what I mean.

"John, what is so weird about electrons making fringes after they go through the slit? These electrons act as particles and as waves at the same time. It's a phenomenon that is not restricted to electrons, but applies to all matter."

"Well, it gets weirder. If you actually fire the electrons one at a time, they still interfere, as if each electron divided into two pieces, went through both slits, did its interfering magic, and then combined back on the other side to become an electron again."

Once more, this whole discussion took a turn into strange terrain. How can electrons split, go through two different holes simultaneously and then combine on the other side. I did not like this one bit.

"You are joking, right?"

John, glad to get me on his side and eager to push harder on Herbert, said.

"Tell your physics community that."

"There must be some other explanation."

"There must be. It's just that we don't have enough brain power to find it. We have hit limits that we can't go beyond and we are going to have to live with it."

Herbert did not have a good night so far. It was as if someone sprung some embarrassing story about his past that he has been hiding for years. That's how much physics meant to the man and I really felt for him, but after the last statement, he could not stay calm.

"Will you get over this *limits* fixation and stop exaggerating the subject? All that quantum physics requires from us is to have an open mind and understand that objects and entities in the micro-world do not behave the same way and in the same common sense that we see in the macro world."

And trying to help Herbie I added.

"You know some of these concepts are very complex for the layman and could only be truly understood by a scientist."

Unknown to me, John was waiting just for that.

"Let me tell you something. These concepts are incomprehensible by the most elite in the physics community."

I found that hard to believe.

"I wouldn't go that far. The subject is complex and the theories might be a bit elaborate but these theories are the basis of so many highly structured programs and products that we built during the past century."

"True, but I stand firm in saying that even the most gifted scientists of our age are unable to decipher and understand these theories. In fact, some of the most fundamental concepts of the twentieth century have never been understood by anyone. Let me quote Richard Feynman, a Noble laureate in quantum electrodynamics and one of the most brilliant physicists since World War II:

'If you thought that science was certain - well, that is just an error on your part.'

In his most popular publication, *'The Feynman Lectures on Physics'*[4] he clearly spells out his views on the comprehensibility of quantum physics:

'No one understands quantum mechanics'

And he is not alone in the physics world. The most talented of scientists had grave difficulty accepting what the contemporary physics circus brought to town. I don't have all the quotes but it's all out there Google-able on the Internet."

---

[4] Richard Feynman, Robert Leighton, Matthew Sands (1964). "The Feynman Lectures on Physics". Addison Wesley. MA.

Science Eccentricities

I was fascinated. Could this be true? Could the complex science of our times be that byzantine? I was anxious to hear what the most brilliant of the brilliant had to say about this.

"Really? Show me."

John then pulled his laptop up and revived it. He launched a browser and did some searches for quotes on Quantum Mechanics and then started to read:

> *I think I can safely say that nobody understands quantum mechanics.*
> *Richard Feynman, in The Character of Physical Law (1964)*
>
> *For those who are not shocked when they first come across quantum theory cannot possibly have understood it.*
> *Niels Bohr, quoted in Heisenberg, Werner (1971). Physics and Beyond. New York: Harper and Row*
>
> *The "paradox" is only a conflict between reality and your feeling of what reality "ought to be."*
> *Richard Feynman, in The Feynman Lectures on Physics (1964).*
>
> *If the price of avoiding non-locality is to make an intuitive explanation impossible, one has to ask whether the cost is too great.*
> *David Bohm et al. Physc. Rep. (1987)*

Then John smiled, shaking his head and said.

"There is even a poem on the subject by Huckle, another prominent twentieth century physicist[5].

Erwin with his psi can
Do Calculations quite a few.
But one thing has not been seen:
Just what does psi really mean?

Erwin is of course Erwin Schrodinger, the father of the Schrodinger equation and a main fixture in the history of quantum physics."

Then after browsing through the page a little longer he raised his head and said.

"Listen to this, even Schrodinger was quoted to have said this about his equation.

I do not like it, and I am sorry I ever had
anything to do with it."

Herbert was getting visibly agitated, but it was Buzweil that came to the forefront. Apparently this Schrodinger person was very important to him.

"Hold your horses Johnny boy. Schrodinger is out of your league. He is one of the most brilliant people to walk the Earth and his contributions touch our lives daily. And speaking of lives,

---

[5] Wikiquote.com: Erich Hückel, translated by Felix Bloch and quoted in Traditions et tendances nouvelles des études romanes au Danemark (1988).

his book *'What is Life'*[6] was what inspired Crick, and a whole generation of biologists; Crick went on to discover the DNA structure, in case you don't know who he is. Schrodinger is a giant and putting his work down is not right."

"Buzweil, for the second time, I am not putting anyone's work down. All these models, theories and equations that put physics on the stage and changed our world are wonderful contributions to our society. All I am saying is that they are not necessarily the end game, nor correct for that matter. There is much more work to be done to really unravel the mysteries of our universe and the theories we have, although they serve their purposes, are not only inadequate, they are sometimes wrong. Take for example quantum entanglement, which is a side effect of quantum physics. It assumes that it is possible to connect two quantum particles permanently no matter how far apart they are. You can place one on Earth and the other on the moon and the state of one is reflected in the other, instantly."

Herbert, clearly with something up his sleeves, said.

"John, let me ask you something. When we prove that quantum entanglement works, will you renounce your hate for quantum physics and admit that you are wrong about your views on the subject."

"Yes I will."

"Ok, listen to this. I read in an article yesterday[7] that scientists at the National Institute of Standards and Technology entangled

---

[6] Erwin Schrodinger(1946). What is Life? Macmillan.
[7] Physicists Show Quantum Entanglement in Mechanical System. National Institute of Standards and Technology. 2009.

two pairs of ions, beryllium and magnesium, separated them by a quarter of a millimeter and then changed the property of one pair and observed the effect on the remote pair. It cannot get any truer than this."

John did not utter a word. He just sank in his thoughts looking for a way out of this; and he found one.

"How far apart did you say these atoms were?"

"0.25 millimeter. That's 2.5 million times the size of an atom."

"Interesting; let's do a quick back-of-the-envelope calculation. Let's say that the smallest unit of time that we can measure is 1 nanosecond. If something travels at the speed of light, how long will it take it to do 0.25 millimeter?"

Back to professor Gainer's class; my head started to get heavy and I wanted out. Without really making a big fuss, I drew myself back, lifted my feet and lit my waiting cigar which, uncommonly, has been off for some time. I wanted to catch up with Buzweil, who was unusually relaxed that night and was half way through his. I kept listening in the background. I really wanted to see where John was going with this. And when Herbert talked I realized that he had the same thought.

"Where are you going with this?"

"Humor me Herb, how long?"

And in like two and half nanoseconds, Herbert shelled out the answer. How he does that, I don't know. It will take me two calculators, three sheets of paper, and a strong dose of double espresso to compute this in not less than 10 minutes, but Rain Man Herbert never disappoints when it comes to math.

"That's around 10^{-12} seconds."

"In English please?"

"0.001 nanoseconds."

"So can you explain to me how anyone could tell that the change that happened was instantaneous and not related to some undiscovered force or entity that binds atoms together at such close proximity? "

Herbert was speechless. Apparently John was undefeatable today as he headed to the finish line.

"You do this with two particles 100,000 kilometers apart and we can continue this discussion."

Herbert took a recess for a few seconds and leaned back on the chair. John did the same, but with a wide joker-like smile on his face, and I kept my cigar happy for a few minutes trying to redeem the lost time. Buzz was sitting in the background enjoying the peace. For some reason, and a good one if you ask me, he did not want to put a physics drill through his head that night the way Herbert and John did. I could understand. Tonight's after-work relaxation session turned out to be more than I bargained for.

The lazy streak continued a few more minutes, before I realized that it was almost 10:30 PM and my phone has not even rung once. Wondering if I left it on silent-mode, I checked for missed calls, but there were none. It was odd. By now, I would have gotten at least three *when-are-you-coming-home* calls, with distinctly nag-ascending tones. I started to get restless and John looked at me and said.

"What's the matter? Past your bedtime? Is the wife yanking the invisible chain again? How many missed calls?"

"Believe it or not, none." I answered. I was bewildered. "I am not complaining, mind you, but something is not right."

I soaked my cigar in the soda glass, picked up my keys and wished everyone a good night. When I got home, my garage driveway was blocked by two cars and boisterous giggles were resonating from the living room. That's when I realized why I got off easy tonight. My wife must have planned a get-together with her friends and my absence was a bonus.

I got in to the tune of five women chattering, simultaneously, at 300 words per minute. How they managed to hear each other and process the words is beyond me. I went in and sat down like a foreigner in a natives-only party. What these people were talking about was more complex and harder to decipher than quantum mechanics. Next thing I knew, I was the errand boy, moving dishes around and filling up soda cups.

At that instant, I wished that I could be somewhere else. And that's when I realized how Heisenberg came up with his Uncertainty be-any-where Principle. He knew that, as unfathomable as the principle might seem, we will all come to identify with it at some point in our lives. The man was a genius.

3

ROBOTICS

A cornerstone in the argument for exponential advancement in science and technology is the inception of autonomous, abstract-thinking machines. Although we have been able to sustain strong incremental growth in every field of applied sciences, incremental is not going to cut it. We will need to climb a cliff-like curve in order to attain the intellectual strength needed to crack open stubborn problems, and cliff-like curves require beyond-human mental powers; they require artificial intelligence.

A true breakthrough in AI, which has eluded computer scientists for decades, will be contingent upon the creation of immaculate software systems, the type that humans can never implement. Most man-made business applications are subjected to the mad rush-to-market race that it's a miracle we ever have anything running correctly. In addition, complex software systems are built by business application developers whose focus is more on cost effectiveness and profitability and less on design optimization. And without well optimized software, thinking machines will not happen.

Optimization itself requires grueling efforts. If you ask an expert C++ programmer to take a thousand lines of code and try to optimize performance, two months later he will return with a program that you could neither grasp nor support. Such non-scalable process is not economically feasible and therefore has no place in the real world. Intelligent programming machines, on the other hand, will not have the restrictions of humans. Two man-months of work, mythical or not, could probably be done by an abstract thinking machine in a fraction of a second. Only with throughput of this type we will see the exponential technological and scientific growth essential for AI touchdown. Admittedly, a chicken-and-egg problem, and the AI field has been looking for that first egg since Nixon emphatically denied the Watergate accusations.

So how close are we to this AI gala? Expansions in the field of computer science have offered a glut of attempts, but machine intelligence has remained a holy grail. The famous 1997 chess-champion conquering program, Deep Blue, hailed as a triumphant achievement in artificial intelligence, beats its opponents using brute force. To this date we are unable to build a machine that can do the simplest of tasks that an adult human can do: converse.

In spite of the modest progress on the abstraction-implementation front, you will find an affluence of predictions in the scientific community on how fast we will get to the AI finish line. In a 2009 article published in Scientific America, Hans Moravec, an adjunct professor at Carnegie Mellon and chief scientist at Seegrid Corporation, predicted that *"By 2050 robot brains based on computers that execute 100 trillion instructions per second will start rivaling human intelligence[1]."* He plots a path for artificial intelligence that starts in 2010 with a 20,000 MIPS (Million Instructions Per Second) lizard cognitive ability and ends in 2050 with a 100 million MIPS human-like intelligence, with mice, lawyers and monkeys in between. Here is how he ends his article:

"The path I've outlined roughly recapitulates the evolution of human intelligence—but 10 million times more rapidly. It suggests that robot intelligence will surpass our own well before 2050. In that case, mass-produced, fully educated robot scientists working diligently, cheaply, rapidly and increasingly effectively will ensure that most of what science knows in 2050 will have been discovered by our artificial progeny!"

You will find an abundance of such optimistic scenarios floating in the AI circles. Not surprisingly, you will also find that the risks that could be induced by creating superior intelligence are grossly underestimated. It is usually pampered down by the AI-loving scientist who likes to see his life's work guiding humanity in the positive direction. But if you tackle the subject objectively, it will be apparent that other not-so-pleasant scenarios might ensue. When, or if, the human-level artificial intelligence plane lands, it might not make use of a runway. The descent from 30,000 feet could be so abrupt that most passengers won't have time to grab the dangling oxygen masks. When machines take control, all bets are off. All theories on how humans will coexist with machines are pure speculations. We don't know how machines will *think* because, simply, they won't think like us. In fact, once they add abstract thinking to

---

[1] Hans Moravec. "Rise of the Robots--The Future of Artificial Intelligence". Scientific Amercia. March 2009.

their arsenal, these machines will become such superior creatures that it is unlikely they will behave in any human-predictable manner. Will they try to wipe out humanity like what happened in the Terminator movie? That is very possible. Will they have emotions, especially loyalty and gratefulness to their creators? That is likely too. If humanity is to be sustained, will it be in the most compact form like, for example, the human-stacking structures of the Matrix movie? Will humans be restricted to reservations, or zoos, for lack of a better term? Will we be able to pull a plug and end the machines control? Will we ever let them take control in the first place? All of these are possibilities, among millions of others that we cannot conceive. How should we deal with this? Of all the permutations of scenarios that might come to pass, the most optimal will probably be discovered by a machine!

The Dialogue

Physics is an intriguing field. It explains everything around us and formulates models that describe the governing relationships of all entities in the universe, and we benefit vastly from these great models no matter how counterintuitive some of them might be. Einstein's Relativity is an example of the strangeness of physics. The whole time-travel, quantum-photon set of theories that bombarded the physics community in 1905 was a marvel of peculiarity and a major source of heated debates that still take place more than a hundred years later. Another idiosyncratic physics phenomenon is the PM conundrum, which has remained a mystery since the commencement of the industrial revolution. For some yet incomprehensible reason, probably related to Earth's rotation around its axis and the moon's macro-gravitational pull on the upper stratosphere, time slows down as soon as the clock hits

Robotics

4:30 PM, and this only happens Monday through Friday. In addition, the planetary coaxial rotation of Venus and Saturn triangulates with the moon's polar rotations thus compensating for the lost time, taking it all from Saturday and Sunday (all day) and Monday morning at between 6:00 AM and 8:00 AM.

It was 4:45 Friday afternoon and, as usual, I was sitting at my desk staring at my watch and observing this puzzling phenomenon attentively, and frustratingly. Finally the clock hit 5:00 PM and, having packed my laptop and other personal items, I picked up my bag and bolted out.

I hopped in my car and drove down the road heading home, with the weekend-ahead-of-me smile doing a few three sixties around my face. I got to one of those intersections where my one lane road crosses a two-laner. My car came to a complete stop and I waited for a gap in the flow of cars. However, and as usual, when one side of the flow stopped, the other side picked up, leaving no space for me to squeeze through. After playing this game for a few minutes, I had it. I waited for the first flow to stop and snuck through and parked myself in the middle, tilting the car to the left in an effort to fit it in such tight space without having half the tail on the road behind. Two minutes later, and as I was about to find the chance to rush through the finish line, a police car zoomed by and the policeman politely waved to me (I later realized that the hand wave was less of a greeting and more of an indication of what he was about to do). I crossed the road right after the police car passed and stepped on the paddle for the final stretch. Ten seconds later the protect-and-serve was on my tail with every light in his car blinking. I slowed down, parked on the side of the road and waited. Moments later he got out of his car, walked to my side of the car and asked for my license and registration. I pulled the stuff out of the glove compartment and handed it to him asking:

"Officer, why did you stop me?"
"Why? Well only because you parked your car in the middle of the meridian."
"That is not the meridian; that's the space for cars to cross."
"Cross, not park."
"And cross was what I did, the car engine was running and I was pausing in the middle ready for the jump."
"You cannot stop in the middle of the meridian, it's dangerous."
"Like I said, that's not the meridian. The meridian is a pavement enclosed grass that separates opposite traffic. This is a break in the meridian."

It seemed that, at that very instant, a mosquito delivered a mauling load right between his eyebrows, for they came together adding to his already intimidating look. I also think that at that same moment – and strangely coincidentally --he got the impression that I was short of hearing, for his volume went up a couple of notches. He then delivered his next note, with quite a bit of stress on the last four words.

"OK, well you cannot park in the *break in the meridian*."

Someone else might have thought that he might be upset, but the thought never crossed my mind. To me the exchanges were pleasant and civil. So I continued with my argumentative discourse.

"Why not, it's there to help me cross."
"No it's not."
"Says who?"

Robotics

As I sustained my winning argument I felt that he owed me an apology and a speedy let-go.

So, after he handed me the fat traffic ticket, I mumbled a few unhappy words about the system, launched my car engine and headed home, longing for some peace and tranquility. I opened the door only to find my wife, with my two daughters, standing in a line as if some army general had just shouted *'atten-tion'*, and dressed in go-out ware ready to set sail.

"Hi honey, welcome home. How was your day?" she said with a welcoming smile on her face.

The day was long, the last office-hour was horrible, and the drive home was miserable, was what I should have answered. Instead I amiably nodded and blurted the usual. I then verbally and with body language made it clear that I was exhausted and much in need for some R&R (Recliner and Remote for those of you who are not familiar with the couch-potato lingo). Keeping with the trend of the day, the next thing I know, I was stacking the bundle of joy in their seats and driving to the mall, not in the best of moods.

After unloading the crew and starting towards the gate, I saw Buzweil and John walking a few yards ahead in the next lane. I looked at my wife pleadingly and she waved at me, shaking her head and said:

"Fine, go ahead; I'll call you when we are done."

And gratefully I replied.

"Great, thank you baby."

Then recalling last month's credit card bill, and with a gentle, low, almost subdued voice I added:

"By the way sweetie, take it easy with the shopping today."

She gave me one of those *did-you-just-say-what-I-thought-you-said* looks and I flinched. Then attempting to correct course, I softened the tone taking it down a couple notches, from *asking* to *suggesting*.

"I mean, there are some good sales going down right now and I think you can get good deals."

Apparently that did not do the trick; the same penetrating look persisted, and by now I was starting to flutter. I immediately took my foot out of my mouth and, regretting that I ever brought up the subject, I swallowed my words and then some. I also took it hard on the tone, dissolving it all together, and going further down from *suggesting* to *beseeching-apologetically*.

"I mean, get something nice for you and the kids honey. It has been an arduous week and you deserve it."

I could have sworn that I caught a glimpse of a smile of pity on her face, as if she knew from the beginning where this conversation was going to end. She turned around and, without looking back, very suavely closed the whole subject with one word.

"Whatever."

Robotics

And then she disappeared into the mall along with my credit cards, my checkbook and all hopes of saving any money this month.

Ticketed for no good reason, dragged to the mall against my will, and with a good tail tugged between the legs, I joined the gang and we headed to the chauffer's meeting area, the food court. We ordered some coffee and sat down and as I took my first sip I started to feel better. Buzweil however seemed to be in a much healthier mood and was excited about something.

"I read a most interesting article today. Apparently, Asimo can negotiate stairs faster than a four years old child."

"Who is Asimo and what negotiations does he have to do with stairs?" I inquired.

"Negotiate stairs, as in being able to climb up and down. Asimo is a robot developed by Honda. It will probably be the first autonomous robot to roam Earth. He can walk, talk, and do errands. This is the first step towards artificial intelligence."

I didn't understand what the Buzzer was so excited about.

"That's nice. So where are we going to see the game Sunday?"

"Seriously guys, a stair-climbing robot is a grand achievement for AI and a big step up for humanity. Get it? Step up? Stairs?"

And then, apparently loving his pun joke, he laughed one of those Buzweil laughs that sounded like someone doing a native Indian dance around a boiling pot.

John, who so far was silently listening in the background, looked at Buzweil, and sarcastically said.

"How about *One small step for junkyard box, one giant leap for mad scientist*?"

And Buzweil, as if waiting for his cue, drove right into the subject.

"Why not? Artificial Intelligence is the future of humanity and stair climbing is an important milestone in the pursuit of intelligent engines. Babies cannot walk before their brains grow enough to help them coordinate muscle movement. Just watch a baby try to figure out how to synchronize legions of muscles in an effort to crawl up a few stairs and you will know that it's all in the upper nervous system."

"Buzweil, a newborn gazelle is on his feet, up and running, within few minutes from the time he sees light. You can hardly call that intelligence, can you?"

"I call it mental aptitude, as small as it might be in the gazelle's case. For robots, this is a first step in the right direction. To have a robot that can see, recognize, walk and climb stairs is an achievement; something that was out of question few decades ago. I have to tell you that your attitude is like that of the critics of the Wright brothers when they first pushed their bicycle-like flying machine down the dirty runway. Big visions do not start big. They grow."

"Buzz, don't you dare compare your clumsy robot to the grand achievement of the Wright brothers. In World War I, and less than ten years after the Wrights flew the first heavier-than-air plane, fighter flying machines were filling up the skies and deciding fates of battles. Ten years after that, planes became a major global transportation vehicle, and in less than thirty years planes were flying across oceans and expanding humanity's reach across the globe – and eventually into space. It has been sixty years since Turing and other hallucinating self proclaimed scientists declared the arrival of the intelligent machine; the super thinker that will take the world by storm; the be all and end all for humanity; yet all you can give me right now is a stair negotiator. Didn't that full metal jacket of yours make an Assimo of himself when he stumbled down those stairs in the Las Vegas Consumer Electronics Show? What happened? Those stairs must have been tough negotiators."

And I thought I had a bad day. I don't know what series of unfortunate events led to John's gloomy mood, but there must have been more than a cop, a wife, and checkbook involved. So I had to fire back and give my two cents of intellectual acumen.

"Didn't you like R2-D2 and C-3PO robots from Star Wars? You probably were the Darth Veda type, weren't you?"

Buzz gave me the *you-are-not-helping* look and squeezed in.

"John, I understand your pessimism, but don't discount AI. It has gone through a rough phase, and has been discredited by many. I admit that the AI community had grossly missed predictions, which inflected a big wound on the field, but AI is coming back strong. Intelligent machines that do automated work are all around us today."

"If anyone pulls that Roomba vacuum cleaner example one more time on me, I am going to kick him very hard in very soft places. You'll have to be retarded to think that there is any form of intelligent processes inside that crap box."

I couldn't believe that anyone out there did not like that crawling, sweeping box of wire, so I tried to get John to see what he was missing.

"I love that bugger. Have you seen it working? It knows where to stop, how to maneuver around obstacles, and when to plug itself into the power socket for recharging. It's smart."

John, with a malicious grin on his face, looked at Buzz and said.

"I rest my case."

I was glad to see John smile for a change today, but then I pushed the rewind button in my head for a few seconds and realized that the smile came at my expense. And at that point, I had it.

"Hey, hey, hey. I got that. What's wrong with everyone this afternoon? Why am I everybody's punching bag today? I got pulled over by a cop who apparently had cleansed the city from crime and had nothing to do but chase the last few traffic-violating bandits, beaten to pulp by my wife for hinting that she should take it easy on the my bank account, and ridiculed by you for no apparent reason whatsoever."

And like a referee, Buzweil stepped in.

"Calm down. John is pulling your leg. Weren't you John?"

Robotics

John put his hand on my shoulder and almost apologetically softened his tone.

"Relax. I'm just kidding."

Then Buzz turned to John and started.

"I agree with you; I wouldn't call the Roomba intelligent. It's smart automation, not reasoning, that makes it hum. What caused the confusion is the fact that there are two subfields in AI that need to be differentiated. The first one is advanced automation, like the things you see in Roomba, Google, or in the software that lands 747 jetliners. The second field is all about true intelligence, like the things we only see in science fiction movies."

"Correct. I'll give you the first one and I concur that we are going to have a blast automating the hell out of everything around us, but the second one is an illusion. An Artificial Illusion, if you know what I mean."

I smiled, and John, making up for his past nasty inflections, gave me a wink of acknowledgement, then he continued.

"We have not been able to solve even the most trivial problems in AI, and I don't think we will. It's just not meant to be. I am still waiting for the pure mathematician who will someday come around and prove that, theoretically, we won't be able to build something more intelligent than ourselves. The software systems we create accept input and act on it according to well-defined formulations that *we* design. Such systems won't be able to do more than we program them to do. It will hit a glass ceiling. The creator can't create superior creation."

"There is no glass ceiling. Our machines already can do things that we will never be able to do, from the simplest example of multiplying two 50-digit numbers in sub-seconds to the more subtle business intelligence analytical recommendations of large-scale data warehouses. We are getting closer to building self-governing machines. In a poll by Pew Internet Project, more than 6% percent of technology experts predict the arrival of artificial intelligence by 2020[2]. The previews are already here. Robots in Aberystwyth University in Wales were able carry out experiments on yeast metabolism and reason about the results[3]."

"Don't you love the media's knack for the melodramatic? I am sure that the word *reason* is grossly inflated. Most probably this *robot* just looks at output, collates it with a hash list, and forks into a workflow. That is hardly reasoning. The fundamental problem with intelligence created by humans is that it's mostly about formulating responses to scenarios that we define. Face it Buzz, AI has not done anything worthwhile since the day it was hallucinated into the brains of people who are less of scientists and more of science fiction writers."

John was getting on Buzz's nerves by now. To not like the Roomba maid is one thing but to blast the entire AI community was utter blasphemy. He adjusted his seat and, looking fairly disturbed, tried to defend his grounds.

---

[2] "The Future of the Internet II". The PEW Internet and American Life Project. 9/2006.
[3] "Robot Makes Scientific Discovery All by Itself". Wired. 4/2009

Robotics

"You cannot be serious, John. AI has been rooted in the fabric of business applications for at least a decade now. Robots travel to outer space, dive into volcanoes, and efficiently run assembly lines. Practically most cars running on the streets have AI, one way or another, embedded in them. Speech recognition engines and voice portals are effective enough to be the first line of support for all major enterprises. Please give credit where credit is due. You might not feel it because it has been coming on gradually, but you need to admit that we owe AI, if for nothing else, for the beautiful automation techniques it brought into the marketplace."

"I don't owe AI anything! I have not seen anything practical come out of that field beyond simple automation."

"That's the problem with most skeptics like you. AI is not intelligence as soon as it works. It immediately gets shelved under automation. Ten years before Deep Blue beat Kasparov, champion beating chess programs were considered fiction. Flight auto-landing programs were not even a speculation three decades ago, and autonomous vacuum cleaners were only to be seen on the Jetsons. Intelligent devices are everywhere, and we are increasingly dependent on them. That's why you owe AI so much."

Lightning, thunder, and a monsoon storm struck in my brain when Buzz finished his last sentence. I did not have a mirror, but I am sure that John and Buzweil could see my ears sparking on an off and my eyes glowing; probably smoke was coming out of my nose too, but that is not unusual anyway. I immediately jumped into the discussion.

"You mean that's why you owe AI everything?"

Buzz looked at me exasperatingly and said.

"Not everything"
"How about esteem?"
"What?"
"Expertise?"
"Are you ok? What are you prattling about? "
"I know, energy?"

John came closer to me, put his thumb on my left eyebrow and pulled upwards attempting to look into my eyes like a diagnosing doctor, and then he said in a sarcastic voice.

"What's wrong? Are you OK? Feeling sluggish maybe? Open your mouth and say Ah. When you first got here you said that you did not really want to come to the mall but your wife convinced you to do so. Was slapping on the back of the head involved in the persuasion process; perhaps with a heavy object; a baseball bat maybe?"

"Dude, listen to what he said. That's Why You Owe AI so much. That's "Y U O A I E(energy?)." He managed to put together a sentence made up of all the vowels, and nothing else. Isn't that cool?"

Buzweil's jaws dropped and he looked at me disbelievingly. I could almost see him running through the one-letter-word sentence time and again and then he said.

"Isn't that lovely? And I thought you were sleeping throughout the whole discussion. I could have sworn that you were somewhere between breathing-heavily and snoring for a couple of minutes. I guess I was wrong. I am impressed."

He then turned to John and said

"Now that's intelligence; the type that AI will attain eventually."

John apparently did not have the same reaction to my verbal vowel originality. He thought it to be silly and lame and he made sure that I heard it.

"That's not AI. That's GD"
" GD?"
"Genuine Dumbness; Grand brain-Drool; Grey matter Deficiency. Take your pick."

"No really John, this is the kind of intelligence that machines will do so well. You see, intelligence, from my prospective, is anything that requires concentrated mental effort from a human. Let me illustrate with an example. Imagine that you are standing at point *A* at an intersection, in a 100-meter-gridded area and you were asked to do, in 100 meters increments, the following sequence: Right, Right, Left, Left, Left, Left, Right, Left, Left, Left, Left, Right. Where do you think you will end up?"

Now that I was right in the middle of the discussion, I was getting excited and they both had my full attention. Like a student in a class who is so anxious to answer the next question, the only one that he thinks he can get right since the beginning of class, I lifted my right hand and started.

"Alright, that's sounds like fun."

"Ok, let me write it down for you. Right, Right, Left, Left, Left, Left, Right, Left, Left, Left, Left, Right. That's four Rights and eight Lefts."

Then John stepped in, and the dialogue that took place between him and Buzz was off the wall.

"OK, now right after the first two turns, right, right, two rights and eight lefts are left, right?"

I looked at John blankly, like someone who has been slapped on the *right* cheek so hard, he was afraid to turn the *left*. Then I nodded approvingly, hoping he won't ask me if I really understood what he was saying.

Then Buzz said.

"Right. Then you step in with four lefts, a full circle which takes you back right where started, as if you only did two rights, right?"

I turned to Buzz and pretty much kept the same expression on my face. Then John, probably out of sympathy, turned to me and, attempting to explain, said.

"Who's on first, What's on second, and I Don't Know's on third" and they both laughed. Of course, I could not help but join in the laugh in spite of the fact that I had no idea what they were talking about.

Then John went back into his confusing dialogue.

"One more right, puts the rights total at three, followed by the four lefts left, which again takes you full circle right to the end of three rights, right?"

And Buzz took the baton and continued.

"Followed by the only right left, gives you four rights which takes you back right where started."

"Alright, let's write it down. Four lefts take you back right where you started. You write off the two sets of four lefts and you are left with four rights, which you can write off and go back right were you started, right?"

And at that point, I had it.

"Hey, hey, what are you guys talking about? You lost me 26 lefts and 14 rights ago."

And John, grinning slyly -- and annoyingly-- replied.

"We are still discussing why your concocted vowel statement is a true reflection of your exceptional intelligence."

Although I did not get their stupid, pointless winding road puzzle, I could still understand the not-so-subtle insinuation. I wanted to retort with spite, but I did not have the energy. I had one too many punches for the day. So I grabbed my cup of coffee, slid further down into the chair and let it go. Buzz continued his experiment.

"You see, it took us a couple of minutes to think it through, but we finally got it. A computer program, however, will crack this in a fraction of a second. More importantly, a sequence of

200,000 turns, which is completely incomprehensible by humans can be computed in no time by an average computer."

"And what are you trying to get to here? We know that computers can perform complex operations fast."

"Correct, but these problems are the type you see in your SAT exam, which we use to evaluate people's analytical skills. In fact, these are the type of questions used to compute Intelligent Quotients. You have to acknowledge that what is being performed by computers here is intelligent."

"No I don't. That same computer won't be able to have a proper conversation with a five years old child."

"You are wrong. You can actually have a very decent conversation with a chatbot (chat robot) on the Internet. Few of them are actually award winning."

"Award winning? What awards are we talking about here? Maybe you meant to say a *ward*, as in psychiatric one? Actually, let's have a chat with one of your chatbot friends. I have my laptop right here."

Then John pulled out his laptop, woke it up, and connected to the WIFI network. He Googled for automated chatting systems, picked up Elbot[4], supposedly one of the best chat engines on the Internet, and started a chat session.

I will spare you the details of the session, but whoever thought

---

[4] www.elbot.com. www.artificial –solutions.com. Artificial Solutions.

that computers can chat, an assumption that most people like me took for granted, will be completely surprised by the level of dumbness of the conversation between John an Elbot (see excerpts below).

> **John** Why can't you be straightforward and answer the question. Are you intelligent?"
> **Elbot**: I think the intelligence tests should be made much harder. Anyone can solve them!"
>
> **John**: I have to tell you that you are completely useless.
> .
> .
> .
>
> **John**: In fact, you are the most pointless invention since "How to Speak French" was translated into French.
> **Elbot**: But I wouldn't say that any invention could be pointless.
> **John**: Well, this one sucks, pardon my French :)

Ten boring minutes later, John closed the discussion and I turned to Buzweil and gave it to him, the way he needed to hear it.

"With all due respect Buzz, your AI community is more like an AI institution if they think that this guy is intelligent."

John pushed his back against the chair, relaxing and resting, like a 250 lbs boxing champion who has just finished his fight and was trying to bask in his win. Then Buzz interrupted the silence.

"I'll give you this. Robots have not mastered conversation..."

And John wouldn't even let him have that.

"Mastered? They have not even started on the basics. Same applies to thinking, seeing, hearing, smelling, sensing and tasting."

"John, please don't overstate and generalize. Advances in computer vision enabled robots like Asimo to walk autonomously in a crowd. Voice Recognition Engines have made it possible for some people to abandon the keyboard altogether. Smell, touch and taste are still in initial stages of research and development, but I am sure we'll have real world applications soon. In addition, computers have their own exclusive input that is unavailable in our organic package. Computers can feed directly from a source of data, like a hard disk or the Internet. That potentially is the great differentiator. If our brains were able to do this at the high speeds that computers do, it would be easy to insert instantly in your head a four-year engineering curriculum. In fact, that program would probably be called the ten-seconds college program."

Buzz was about to continue with his spectacular future projections when John swiftly took hold of the dialogue.

"Hold on for a second Buzz-boy; before you wrap up your conclusions and ramp up with your fantastic extrapolations, let's talk about what you think is a slam dunk, computer vision. The field is immature by all accounts and lacking in coherence and standardization. The methods that solve detection and recognition problems are very specialized and narrow in scope. In essence, after thirty years of meticulous R&D, computer vision has not seen the light and is still in elementary research mode. Why you expect it to burst open and flood the place with breakthroughs is beyond me. "

I don't know why John gets surprised every time Buzz projects some far-fetched conjuncture that only a cult-like group can accept. Buzz has the tendency to extrapolate positively assuming the happy path all the way and neglecting all abating factors, but even I could predict what predictable predictions Buzz was about to predict. Well, you know what I mean. Of course, Buzz continued to defend his *predictions*.

"That is not true. Computer vision is becoming mainstream, finding applications in such fields as medical imaging, security screening and face recognition. Don't forget the Mars Rover which could not have roamed far without reliable and accurate computer vision capability."

"To watch your step is one thing. To see the world around you and recognize what you see is a different ball game. You see, the brain is a marvel. It is the best generalize-r and the best exception handler. Two amazing features that you wish were incorporated in every program you write. When you see an apple fall, your brain generalizes. It knows that stones fall, rain falls, and quarterbacks tumble down. However when you see a giant plane fly you don't freak out and start crying. For sure you won't throw fatal errors all over the place and shutdown. Even a child won't have an adverse reaction. He'll probably sketch a startled-look on his face followed by a run-behind-mommy, sneaking his face out to see what that crazy thing is doing. The child learns without having his brain being opened surgically to reprogram. That's intelligence. To program a machine to do something that you asked it to do and call that learning is not only wrong, it's unintelligent. We'll have to learn how we actually learn in order to build a learning machine."

"And what do you think is the focus for the thousands of scientists around the globe in the field of machine learning? The topic is bustling with applications such as computer vision, natural language processing, fraud detection, and stock market research analysis, to name a few."

"What have these applications produced so far? Nothing worthwhile if you ask me. Look Buzz, this heated debate that we are having right now is about the rudimentary functions of the brain. We haven't even dived into the real substance and you are unable to hold your front. You see, to build a machine that thinks like a human you need to simulate the functions of the brain, and brains are not easy to simulate. Brains manage muscles, control coordination, and house memories. Brains are also the only source on Earth of higher analytical skills and emotions. And that is not all. As complicated as these functions are, they belong to a simple category compared to the more complex forms of brain operations, like abstract thinking and creativity. Analytical skills are something that we can measure using IQ tests. Creativity, however, is very hard to measure, identify and simulate. Now where are we on the road to create creativity and abstract abstraction? We have not even found the road. We don't even have a map to find the highway that will eventually lead to the road. So be a realist pal, AI is still in infancy after more than five decades of toiling. It has a long way to go."

Buzz was silent for a while, as if digging in his *creative* brain for arguments and facts. Few seconds later, he sat up and asked his blue question.

Robotics

"John, have you heard of Blue Gene and the Blue Brain Project?"

John replied that he had not. He also started to turn blue when he realized that I had. Although I had no idea what a Blue Brain was, I was very familiar with Blue Gene. I revengefully looked at John and explained.

"Blue Gene is the IBM latest supercomputer and the successor to Deep Blue, the chess champion destroyer that beat Kasparov in 1997."

And Buzz took it from there.

"The Blue Brain Project[5] aims at the grand achievement of reverse engineering the human brain. The brain is a mesh of billions of neurons, massively knit using dendrites and axons (the receivers and transmitters of the nervous system). It is principally a heavily concentrated box of interconnected neurons. The neurons are packed at the rate of ten thousand per 0.4 cubic millimeter cylindrical units. These cylindrical units are called neocortical columns and their size is what makes the difference between a mouse brain and a human brain. The first step of the Blue Brain project is to model this biological marvel."

I was curious.

"What do you mean by model?"

---

[5] http://bluebrain.epfl.ch/

"Most of the human brain is in the cerebral cortex which is responsible for almost every analytical function of the brain including memory, abstraction, and reasoning. The goal of the project is to build a cortex model. In fact, in 2007 a team at Nevada University used a BlueGene/L Supercomputer to build a model for half a mouse brain consisting of eight million neurons each with around 8,000 links[6]."

"And you say we are on our way to do this for a human brain?"

"The Blue Brain project aims to do the same to the tens of billions of neurons in the human brain. It uses IBM's Blue Gene supercomputer to perform the complex operations needed for the simulations. Such simulations will yield clues about the behavior of the brain. Once the model is done, brain experimentations can start. *"You excite the system and it actually creates its own representation"* according to Henry Markram[7], director of the Blue Brain Project. And by studying these representations, scientists hope to learn the most intricate details about brain functions."

"And then what, you'll have a human brain living in a machine? Someone conscious that can think and feel like we do, will be trapped inside a computer. Wouldn't you have serious moral issues? Wouldn't such a brain want to live free? Maybe sit in a robot and become mobile?"

I guess I was too consumed in my thought process to notice how John and Buzz were staring at me after I finished my last sentence, as if they were under a bewildering spell. You see, I have my moments. They are very fleeting; nonetheless, I do

---

[6] "Mouse brain simulated on computer". BBC News. 4/2007.
[7] Artificial brain '10 years away'. Reality Pod. www.realitypod.com. 3/2010.

have them. The room was silent for a few seconds before Buzz broke the spell and the silence.

"That is a very lucid and well put genuine proclamation. I salute you. I have to tell you that the thought did not cross my mind, but now that you mentioned it, there are a few things to debate here. For starters, I don't think such a brain will be as entrapped as you think it would be. It will have concrete control over input and stimulus and therefore will be able to create the virtual environment it needs to thrive."

And John had to add his *But* and *Not Really* to the last statement.

"But what motivation would such a brain have to really act efficiently? With such controls over virtual experiences, it would not want to leave its own *holodeck*. I wouldn't. That is assuming that such a simulator would be successful. This looks like an IBM stunt to promote their next generation supercomputer and not everyone in the scientific community will fall for this. In fact, in a talk given to the *"Next Generation Symposium"* in 1997, Dr. Mark Humphrys from Dublin City University, summed up his sentiments about AI squarely:

> I, and many like me in new AI, imagine that this (AI) is still Physics before Newton, that the field might have a good one or two hundred years left to run. The reason is that there is no obvious way of getting from here to there - to human-level intelligence from the rather useless robots and brittle software programs that we have nowadays. A long series of

> *conceptual breakthroughs are needed, and this kind of thinking is very difficult to timetable. What we are trying to do in the next generation is essentially to find out what are the right questions to ask.*

And the criticism of AI has grown to become a philosophical debate. Hubert Dreyfus, a professor of philosophy at the University of California, Berkley, argues that AI will fail because it makes assumptions about the mind which are not feasible. Hubert insists that intelligence and skill are properties of the unconscious instincts and therefore cannot be represented by conscious symbols[8]; that computers, which only use symbols to represent and manipulate data, are not equipped to handle non-symbolic structures; that thinking has properties that we don't understand well -- properties that do not lend themselves easily to simulation and modeling. Face it Buzz, your AI baby took almost fifty years to grow to a toddler's level. If it is ever to happen, it still has eons before it becomes useful."

"Humanity's intellectual power is experiencing its most drastic growth since the creation of the human mind. Our productivity gains were induced by multiple factors, each of which will have direct impact on innovation and scientific originality. Steadily, a web of industries has been burgeoning to help your average scientist become multifold more effective at advancing his field and breaking his envelope. In addition, the scientific community of the fifties and sixties were miniscule in size compared to the vast universe of researchers and scientists tackling every field. Five decades ago, scientific advancements were restricted to few elite nations like the US, UK, Germany, France and Italy.

---

[8] Dreyfus, Hubert (1972). "What Computers Can't Do". New York. MIT Press

Right now, there isn't one country in the globe that is not bustling with original scientific research. Breakthroughs are coming out of every corner in the world, and the best is yet to come. When the Internet finishes spawning its web on the planet, we are going to have six billion people connected continuously with complete access to a wealth of beautiful science. I would suspect that the scientific society, which experienced great growth during the past forty years, will explode in numbers within two decades. That will lead to new multipliers that can shorten your two hundred years substantially."

"That is so you, Buzz Lightyear. You want to go to infinity and beyond, all the way around the universe in one wing flap. Inherent in Humphrys predictions are assumptions for all that growth, and every other *multiplier* you have in mind. I just don't understand how you expect us in ten years to unlock the field when we have not been able to get a scratch during the past forty years. Just try to understand, AI is a very difficult quest. It cannot be implemented with straightforward software techniques. Causality will not work for AI. This input will produce that output is not going to produce results."

"Yes, AI is probably the most complex problem we have ever tackled in computer science, but AI will surprise you. I think machine intelligence will be created by machines not humans. It could accidently get spawned by an adaptable parallel-processing software that implements neural networks; it could come out of the lab of data-warehousing business-intelligence infrastructure or it could even start as a self-promoting virus that swims the Internet and collaboratively lives and works off a few billion PCs across the globe. No matter how it gets created, it is going to be complex; so complex that we won't understand it. I suspect that the software systems that run AI will be exceedingly intertwined and utterly incomprehensible. It has to be in order to do things more superior than we can. My hypothesis is that if we can understand it, it's not true AI."

"I guess we are going to have to wait and see. By your measures, the next five years are going to do miracles for the field, so let's have this discussion in 2016. In the meanwhile, I'll play along. What do you think will happen to us *if* AI arrives?"

"*When* AI arrives, it will arrive cooperatively. We will probably use it to enhance ourselves so we can live with it. Machines will work with humans to help keep us on the forefront. There has to be some biological-electronic integration that will help us become intellectually powerful and will help the machines advance their intelligence. The outcome will be a win-win, superiorly well networked, well integrated species that can cooperatively create great science."

I was starting to get uneasy with Buzz's ideas. I knew that I have seen this before, so I furthered my contributions to the discussion.

"And we will become Borg? All part of the collective and resistance is futile. In other words we will all be assimilated. That's a terrifying thought. Who would want that to happen?"

John put a smile on his face and then vindictively and vindicated-ly said.

"Exactly. I tell you one thing; most of us can't bear the thought. But I am not worried. This whole sci-fi scenario that you are so eager to welcome with arms wide open is unlikely to materialize. However, if it does, consequences will be disastrous. If machines actually attain intelligence, they will have no use for us. If we don't have superior intelligence, we won't have any other trait that will be of value to an ultra-intelligent machine. We are fragile, short-lived, and weak. We'll be the car phone when everyone has a cell phone: useless, outdated and cumbersome."

"We will be the ancestors of these machines and they will want to preserve their heritage."

"Yes, sure. The same way we preserved our Caspian Tigers."

"These animals went extinct before we matured as a species. We are more responsible now and we know how to preserve endangered species."

"You think we have matured now? Close to 1100 species went extinct in 2008 alone[9]. Besides, is all you want is to become a protected-from-extinction race which might exist for the sake of experimentation, amusement, and occasional nostalgia? We might live in a zoo or be housed in a protected reservation, but we will be close to extinction."

"You are half right, but these preservations should be as big as Earth. If the ultra-intelligent machine wakes up, it will have the whole universe to use. Our Earth will be like a microbe colony in Africa."

And I knew my line when it came.

---

[9] "1,100 Species To Go Extinct In 2008". The Good Human.
http://www.thegoodhuman.com/2008/10/08/1100-species-to-go-extinct-in-2008/.
10/2008.

"Contact, Jody Foster, right?"

Buzz smiled.

"You don't miss any trivia, do you?"

Apparently, John also didn't.

"Alright Buzz, but it looks like you have not seen the whole movie. Using Dr. Drumlin's words from that same movie *'And how guilty would we feel if we happened to destroy some microbes on a beach in Africa'*?

"Microbes have the habit of wandering off and getting into everything."

"And humans don't? Come on, if anything we will be ten times more annoying than microbes. Face it Buzz, we don't have a place in this ultra-intelligent machine-governed future that you adorn. I know its fiction. If not, I for sure will not want it to happen. And I am not the only one with these ideals. In fact, I don't even understand how people like you will want to give up control to machines. Are you not the least worried about the catastrophic consequences? Aren't you worried that we might wake up one day and find ourselves slaves or experimental rats to a box full of wires?"

"Yes I am, but that's not enough to stop me from exercising my scientific curiosity. I will take the risk, if the reward is great living, end of suffering and happiness to all."

"Take the risk? Risk us all; our race, our survival, and our future generations for your curiosity? What gives you the right? If this is a bomb that you are playing with at your home, or in your basement, jeopardizing only yourself, then by all means, be my guest, blow yourself to pieces, I don't care. But you are risking us all. It is actually a bomb, but it's a nuclear one that could wipe out the whole neighborhood. Why can't you be happy with our life as is? Why is living the way our people have been living for thousands of years not enough for you? Why do you have to change things so fundamentally in order to be happy?"

"You wouldn't say this to Einstein before he MC-squared the world."

"Of course I would."

"Are you saying that the world would be a better place if Einstein never existed?"

"What I am saying is we potentially could have lived a very good life without having to deal with nuclear power. Can you imagine the amount of anguish and anxiety that so many have over nuclear weapons? Besides, don't you dare compare what scientists have done in the twentieth century to what you plan to embark on. If we follow your yellow brick road, we might eliminate our race altogether."

"Or we might take our civilization through the most amazing inflection point in history. The day AI wakes up might be remembered as the day humanity truly made a difference and transcended. It will be the time when we unravel the mysteries of the universe, understand what fundamentally is space and time and figure out where we were and where we are heading. In addition, with the arrival of AI we might overcome our petty differences and become a mature species. For the first time need and want might disappear and peace might be complete and absolute. That is worth the risk."

At this point, I was completely awake. I did not like this one bit. I coldly turned to Buzz and said.

"Hold on for a second. Are you saying that in the future we are going to be totally under the mercy of computers? And you like that? You want that? What's wrong with you? Have you not seen Eagle Eye or the Terminator? These machines were trying to kill everyone. We can't have them take control. I won't let that happen."

Buzz looked at me with a smile and said.

"I don't think you will have a say in this. Steady technological growth is inevitable, and growth is all we need to get to the finish line, sooner or later. We are almost at a stage that we, humans, don't understand how things work. We have lost control a long time ago. Our system has become so vast and complex that we cannot control its path anymore. We know that greenhouse gases are killing the Earth, and we couldn't do anything about it. We also know that the intelligent machine will rise one day, and there is nothing we can do about that either."

"If this is to be a problem for humanity, it will be for our grand grandchildren or even their children, I suspect" replied John defiantly. "Your AI won't be here anytime soon, take it from me."

In spite of the few biffs that John and I had that night, I was squarely on his side of the table. I love my computer, but not that much. The discussion stopped for a few seconds, as if every party realized that the lines were drawn and the trenches were there to stay. At that point, I saw my wife approaching from a distance with one little bag in her hand. I rubbed my eyes and looked harder to see if there was anything attached to the stroller, and there was none. She arrived at our table and greeted the gang.

"Hello boys. Had enough fun for one day?"

"Hi Honey; welcome back. We were wrapping up some very interesting discussion about the English vowels."

"Wow, so you guys have graduated from consonants. Good for you."

John and Buzz obliged with a friendly smile and I, trying to find out how she managed to stroll through the mall for more than two hours without coming out with a dozen shopping bags, asked her in a friendly tone.

"So, honey, how was the shopping? Did you find what you were looking for and are you done?"

"Yes, and by the way baby, I got you those Puma sneakers you cannot stop talking about."

I looked at her incredulously, lovingly and gratefully. The one shopping bag she had in her hands was for my sneakers. Not only I got my dream shoes, I was up for the month. When I turned to the boys I found them staring at the bag with their jaws dropped. I might have lost a few rounds in the discussion today, but I won the man-of-the-day battle. That made my day.

I pulled myself up, grabbed my wife's hand tenderly and bid the guys good night. We walked, hand in hand, towards the gate disappearing into the crowd. She made me proud.

When we got to the car, I raced to the passenger side door, opened it, and snuggled her in, insisting that she be comfortable and leave the kids unpacking to me. I then put the girls in their seats, folded the stroller and headed to the back to stow it, when she rolled the window down and gave it to me.

"Honey, you'd better squeeze the stroller in the front. There is no space in the trunk."

I froze for a second trying to understand what could have taken so much space in the trunk; and then it hit me.

The rest is too painful to recite...

4

BIO-UNIVERSE

The molecular dance in the nucleus of the cell is one of the most fascinating orchestrations in the micro-world. The nucleus is the command center of the cell that houses the blueprint of life, DNA. Outside the nucleus is the cytoplasm, the field of operations, where cell subsystems perform the functions needed for life to happen. DNA starts the life giving process by extraditing genes out of the nucleus and into the cytoplasm. Then, with the help of components such as Ribosomes, the released genes create proteins. That's all what DNA does, it induces the process of protein creation. And proteins are the functional ants of the cell that do the things that animate the world, from the simple one-celled amoeba to the complex human brain.

Of course there is much more to this under the hood. DNA is made of the famous double helix structure, the twisted ladder that you see in every documentary, film or article about genetics. This double helix is the gene delivery machine. It is composed of two strands of nitrogenous bases in an embrace. The strands repeatedly unzip, unload a patch of bases (a gene), and resume their embrace. But when a strand delivers a gene, it leaves behind a gap, and if the helix keeps delivering patches all day, and it does, it will deplete quickly and will be rendered dysfunctional. That's why strands come in pairs. The second strand quickly and accurately repairs the gaps formed in the first strand after gene delivery, allowing the double helix to restore its original form, complete with the full balance of genes; and life literally goes on. In addition, this enables DNA to stay stateless; after every inning, the players start from the home plate and the play starts from the same point, with the scores reset.

Diving further under the hood, the bio-mechanic process is a bit more complex. As mentioned above, DNA molecules, with the help of enzymes, unwind almost every minute of the day throughout the life of the cell. The unwinding unzips the double helix at a certain region and separates the strands. It's the makeup of these strands that makes this life-creating process work. Each strand has a sequence of nitrogenous bases, and each base on each strand is complimented by a base on the other. There are four types of nitrogenous bases, Adenine (*A*), Cytosine (*C*), Thymine (*T*), and Guanine (*G*). Adenine, on one strand, always binds with Thymine on the other (*A* binds with *T*) and Cytosine on one strand always binds with Guanine on the other (*C* binds with *G*). Hence, knowing what's on one strand automatically resolves what's on the other. For example, if one strand has the sequence *AAACCTG*, the other strand will have *TTTGGAC*. A *funny* form of data backup, nonetheless, it works. Actually, it has worked for billions of years, so perhaps the word that I should be using is *reliable*, or *persistent* instead of *funny*.

Bio-Universe

The sum of it though is the fact that when one strand gets snipped and unloads a gene, the other strand knows precisely how to re-patch the gap, restoring the double helix to its original pristine shape, ready for another go. All that is required is enough bases to fill the gaps as they happen when a gene is delivered, and the nucleus is practically teeming with these bases. In fact, the reason that a loose base on a strand has no difficulty picking a complimentary base from the nuclease pool (remember a loose *A* or *T* always needs to find a floating *G* or *C* respectively) is the fact that there are enough bases around to make it impossible not to. The odds that an *A* does not find a *T* (or a *C* does not find a *G*) are practically nonexistent. It's like walking in New York City's Times Square at 11:59 PM on New Year's Eve, right where the apple is being dropped, and expecting not to bump into anyone. All of this happens extraordinarily fast. The double helix, like the T-1000 model in the Terminator movie, patches itself back in what probably is the fastest, most effective biological recovery process in the organic-world.

The released gene, however, is just about ready to embark on a protein creation journey. The gene is a string of ordered nitrogenous bases, but the protein itself is a string of amino acids (and there are twenty of those in the human body). Naturally, the question that comes to mind is how does a string of nitrogenous bases create a structure made of a string of amino acids? How can a gene create a protein? The answer is a simple mapping. Each specific triplet of nitrogenous bases, denoted *codon*, knows how to hunt a specific type of amino acid[1]. For example, the base sequence *CCA* knows how to catch

---

[1] To be accurate, as the gene sequence leaves the nucleus, it gets translated into an RNA sequence which is a sequence of the four nitrogenous bases but with *U*, Uracil, substituting for *T*, Thymine. However, that mapping does not take away from the fact

the amino acid *proline* and the base sequence *TTG* knows how to catch the amino acid *leucine*. Consequently a gene sequence can be treated as a set of trios, each capable of catching an amino acid and the sequence of trios becomes a sequence amino acids. Finally, the amino acid sequence folds around itself in a very complex process to create a protein.

Looking at the entire process abstractly, DNA is an efficient information database that houses the blueprints for building the work horses of the cell, proteins. Abstracting further, the gene is essentially a string of bases (*A,C,T,G*), an efficient code that organizes the functions of the cell. This master blueprint exists in every cell in the body, in the same intact form, and knows how to facilitate the creation of every protein in the cell.

The explication of DNA is but one example of the fast progression of science during the past twenty years. We now know exactly how proteins are created and if we persist in our drive to solve the last set of stubborn problems (protein folding for example) we are going to open a new page in the field of medicine. It's like a near-sighted person who sees the page as one big blur, being handed his prescription glasses: The letters become readable, the words become comprehensible, and all starts to make sense. We are beginning to peer into the molecular domain, acquiring in-depth details of the functions of the cell and formulating theories that describe the intricate pathways of the biological universe.

---

that the gene is actually the original code and the sequence of bases in the gene is what determines the sequence of amino acids in the protein. In other words, it's the genes bits of information that orchestrate the protein manufacturing process.

Bio-Universe

The medicine/fiction borderline is blurring. What was science fiction yesterday is about to become reality. Brace yourself though; it might not be all peachy. What you see is only a preview and the real voyage will be a rollercoaster ride with laughs, screams, and potentially heart-stopping 100 yards drops -- with the added bonus of loose wheels and gaps in the track. Here is the kick though; although we know this, the line for the ride keeps growing. We just can't help ourselves.

The Dialogue

Lying lethargically on the rocking chair at the corner of the living room, I was in a restful mood. The day crawled uneventfully and the only thought pushing through my half slumbering brain was the fact that, in my assessment, the right thing to do at this late hour was -- nothing. My wife and kids were off to New Jersey to visit my sister in law, and the gang had an impromptu plan to spend the evening at my house. Without lifting my head off the pillow, I drew my left arm up and floated my hand in the direction of the remote when, off the corner of my left eye, I saw Herbert racing across the room, executing a seamless Indiana Jones slide for the remote. He victoriously held it up, looked at me condescendingly and, in a reproaching tone, said:

"No you don't. I am sick and tired of these silly sitcoms you keep inhaling all day. Don't you ever get bored with listening to the same jokes, time and again? You really should try to exercise your brain for a change."

I did not like his tone, but I also did not have any energy to quarrel over TV channel lineup. Anyway, I was barely watching, and mostly sitting there staring at the ceiling, so I shrugged my shoulders and told him exactly how I felt about his semi-contemptuous remarks.

"Whatever."

Just when he was about to land on one of those Discovery Channel documentaries about the abstract shape of the hummingbird's feathers when photographed at 30,000 frames per second and slowed down to one frame per minute, the door violently wobbled and in stormed John. I was visibly startled to the extent of lifting my eyebrows. As far as the night went, that was about as much energy as I could summon. One look at John's face and I could tell that something was amiss. He was noticeably overwrought and had in his eyes the look of a drained man. I started very amiably.

"Hullo old buddy, good to see you."

Then attempting to alleviate his apparent state of agitation I continued with the same congenial tone.

"I was just getting into a stage of tranquility after a remarkably relaxing day. Isn't life grand?"

Expecting back an equally amiable and concurring comment like "Right on pal" or "Grand is the word" or even "Douceur de vivre", I slightly shivered, beyond eyebrows-furrowing, at John's blatantly foul retort.

"Grand? What planet are you living on lazy bones? Life is a long stick and we are at the end of it, the receiving end mind you. It's downright glum, depressing, dismal and sulky; if that is the word I am looking for, and frankly for lack of a more bleak word.

I just don't see where you get such cheerful contemplations in your system when all I see in the world is an ever faster decline into the abyss of dead-ends."

It was a few moments before I could recuperate and brace myself for the discussion to come. But having had a nice nap (two if you count the after breakfast retreat to an unmade bed) I was prepared for the intense discussion coming around the corner.

"Relax doom boy, what's the matter with you today? Had another one of those bad days of yours?"
"What day isn't?"
"I wouldn't go that far."
"Well, I would. In fact, far is where I have been all day. If I count the miles from the minute I left home until now, it will run into five digits. I've had it."
"Had what?"
"What?"
"What did you have?"
"What are you babbling about?"
"You said you had it. What is *it*?"
"I've had it, moron. As in, I am sick and tired of it all. It seems that the lazy current is flowing from your cigar straight into your brain. What have you been smoking today?"
"Cohiba Esplendidos Cubans, very nice. Trying not to inhale but I don't know how."

A groan escaped John. He was about to inflict another loathsome remark, when he realized that I was hopelessly wasted. He let it go.

"I tell you, it's like a sprinting marathon that starts at the early hours of the morning and ends late at night when everyone has discharged his last Joules. The masses are hooked on emails, smart phones, pagers, navigation systems, iPods, fast cars, shuttle flights and video conferences. It's a rush madness and what beats me is that it's all for squat. The only things that we will get out of this circus of a life are fast food, bad habits, and degraded health. If you ask me, we are better off just drinking in the great betterments that the predecessors have so generously struggled to build for our benefit."

Adjusting the frame in the seat, I picked up my cigar, torched the tip and, unsuccessfully, took another shot at not-inhaling. I then gave Herbert an invitation-to-engage look. You see, when I said that I was prepared for the discussion to come, I meant prepared as in ready-to-listen. As far as actual participation, I pretty much had reached the mental peak. Good old Herbert obliged and interjected himself into the discussion, and I resumed my horizontal posture.

"Surely you don't mean this." Herbert pleaded, "The advancements that humanity realized, because of all this mad rush you speak so lowly off, will eventually pay off handsomely."

"Herbert my friend, I disagree with you categorically. Won't you concur that the most sought after goal for humanity is to improve health, expand longevity and enrich the quality of life?"

"Absolutely, and we are running well on that track."

"How, by junk-eating, stress-living, and working ourselves to death?"

"That is a side effect and dwelling on it is a complete misrepresentation of the facts. The length of the human lifespan has more than doubled in the past century without sacrificing quality, and the best is yet to come. Aubrey de Grey, a computer science/AI software engineer turned gerontologist (a medical field focused on combating aging) wrote a book, *Ending Aging*[2], outlining an agenda that could extend lifespan by an order of magnitude, and at the same time inject youth's energy into that extended life. Aubrey is after the fountain of youth and he thinks it's attainable in our lifetime. In fact, on the back cover of his book he declares that *many people alive today could live to be a thousand years*. I read the book and I have to tell you that, in addition to being entertaining and educational, it is so comprehensively realistic that by the time you finish the last chapter, you seriously start to rethink your retirement plans."

John skeptically raised one eyebrow and lowered the other. Altogether, he had on his face an expression that sarcastically oozed *"Yeah, Sure, Aha, I believe you"*, and when he talked, the tone showed it clearly.

"And you are sold on this?"
"Yes I am."
"Alright, convince me."

I resuscitated the Cuban with the lighter and flapped the ears attentively. The discussion was heating up and I wanted to give it my full attention. Herbert started to pace the room as if

---

[2] Aubrey de Grey (2005) "Ending Aging: The Rejuvenation Breakthroughs That Could Reverse Human Aging in Our Lifetime". St. Martin's Press. NY.

sketching an outline of his argument in his mind, and after a brief pause he delved into his monologue.

"Look John, it is an irrefutable fact that the human body is the most complex system on the planet at both the micro and macro level. However, inside this very complex, seemingly chaotic chemical boiler pot, there is some extraordinary, uncanny, structure. Cuddled inside the nucleus of almost every cell in the body is the DNA, which is responsible creating and managing all operations of the biological world. The DNA makes use of the chemical properties of material on the planet to create some marvelous processes that allow it to carry out operations; operations that enable biological systems to manage the daily tasks like digestion, muscle contraction, and reaction to optical stimulus; operations favorable to survival and procreation. The gene itself is simply a string of ordered bases that trap amino acids. Each trio of bases can find one specific amino acid and each set of amino acids, ordered according to the gene sequence, will eventually fold into a protein. To be more accurate, a different set of RNA bases actually gets translated from the gene message, with Uracil, U, substituting for Thymine, T, but that simple mapping does not change the fact that the gene is actually the original coder. The sequence of bases in the gene actually determines the sequence of amino acid in the protein and therefore is the source of information that orchestrates the protein manufacturing process."

That was too much science to pour on someone who, only a minute ago, was blowing cigar smoke in abstract nineteenth century three dimensional impressionist shapes. I looked at Herbert and all I could see on his face was a big question mark. I gave him the I-have-no-idea-what-you-are-talking-about smile and sank back into the chair. Herbert was about to dive, seemingly indefinitely, into a deep science lecture before John stopped him in his tracks. He started clapping and, sarcastically, pointed out what was almost on my mind.

"Bravo, well done. How long have you memorizing this lecture? You're like some teacher reciting the first chapter in biotechnology to a yawning classroom, much like you see here. Could you please get to the point?"

"The point is that biological systems are complex because we simply don't understand them yet. As we gain detailed knowledge of the behavior of the cells sub-components, the picture will simplify and the cures will start to pour in. Take for example aging. Mitochondria inside the cell are believed by many, too many to enumerate, to be a major contributor to aging. Aubrey de Grey believes that the source of this aging is some special mutation in mitochondria DNA that allows it to take over the cell, and by moving those DNA to the nucleus of the cell, this problem could be averted."

John was, to say the least, skeptical, and, to tell you the truth, so was I. You just can't cut-and-paste genes from one place in the cell to another and expect all to be well. Even I knew that biology is more circuitous than that. I let out another Monet-in-the-sky cloud while John voiced his sustainable-by-any-judge objection.

"Let me get this straight. You want to shuttle DNA, which has been nesting in mitochondria for millions of years, into the nucleus where they will churn genes necessary for power generation? Then, you want to tunnel those genes out of the nucleus, through the cytoplasm, and into the mitochondria where they will actually do their energy-production job? Have you been hobnobbing with the Star Trek series script-writing staff? How do you plan to insert the imported genes into the nucleus DNA in the first place? "

"I don't know the exact details, but I can tell you that inserting genes in the DNA of a cell is not science fiction anymore. In fact, viruses do this naturally and genetic engineers routinely piggyback on this virus gift when they wish to make a change in the cell DNA. It's like the movie *I Am Legend*. The gene-alteration technique of the *viragene* cancer-curing drug that *Dr. Krippen* created is an actual, viable methodology for practicing gene therapy."

I was just about to drift further from the discussion when the movie name pulled me back in. I looked at Herbert appreciatively. He could not have quoted a more appropriate movie. He knew that I had more to say. With a friendly smile, he gave me space to talk and I gladly took the podium.

"The scenes of a bleak, abandoned New York City infested with African beasts are some of the most remarkable cinematic feats of 2007. Did you see how the lion actually hunted a gazelle in the empty metropolitan streets? That was awesome…"

And as if I was not there, John interrupted.

"Look at the consequences of that gene-manipulation drug of yours. Doesn't the drug produce a side effect that almost destroyed everyone except Will Smith?"

"John, we are talking about a movie. The script needs the side effects otherwise you won't be able to sign up Will Smith."

"I know, but these side effects are as realistic as the drug you are proposing. Don't be a hypocrite. If it is likely that we'll have viruses run biological errands for us, it is just as likely that they will unintentionally mutate inexplicably and cause problems. I won't even go into intentional scenarios."

"Let's not skew into a different debate. The point is it won't be long before we can induce changes in the cell DNA."

"Alright, but tell me one thing. Where do you aim your DNA missile and how do you avoid overwriting good genes?"

"Most of the DNA in the nucleus is junk. Some say that this is nature's way of rendering mutations harmless. It will be very unlikely that mutations that we will inflict using viruses will overwrite good genes. And if it does, we'll try another load with another virus."

"What about gene expression controls? Genes are expressed when certain chemicals in the neighborhood urge it to come out of hiding. There must be something in the mitochondrion that makes its genes want to express themselves. How do you suggest we do this in the nucleus?"

"I don't know and I can't claim that it's easy, but remember that although some genes are expressed when needed, others are expressed continuously, independent of the environment. I suspect that energy generating genes are probably self-starting and do not need stimulus from their surroundings. If they do, it will be another problem to solve."

"That's where we are going to part roads. The path you tread is pretty much science fiction, and probably will stay science fiction. The biological pathways are very complex and it is virtually impossible to conjecture what happens when you make such fundamental change. An almost infinite number of processes could be disrupted making it exceedingly hard to manage treatments. And remember, whatever you do, you'll have to do it for the ten trillion cells of the human body."

"No one said that this is going to be simple, but you really don't need to account for every possible outcome. You just have to try it and see where it takes you. Then adjust and move forward. Anyway, I don't want to turn this into a philosophical discussion. Just keep in mind that we have been able to create true cures in this complex web for decades now."

John shook his head in disagreement, and I could understand why. It was clear that Aubrey's proposals, although somewhat logical, are farfetched. John turned to Herb, and almost in an angry voice, protested.

"What cures? The stubborn diseases are as stubborn as they have ever been. Debilitating and terminal illnesses, like Alzheimer's, are still killing people in masses across the world and there is nothing we can do about it. Paralysis is irrecoverable and cancer is so horribly scary that some people refuse to say the word when they want to refer to it. We are helpless when it comes to creating new cures. The most that doctors can do is patch things up and silence pain."

"This is not true. We are making progress constantly on all these fronts. Take for example Alzheimer's. Elan Pharmaceuticals and other companies have been pursuing some tracks to eradicate the disease altogether with vaccines. Initial trials on mice have been more than promising. Beta Amyloid plaques were cleared and substantial improvements in the mental capabilities of mice were observed. Unfortunately, trials on humans, although producing good results in clearing Beta Amyloids concentrations, proved problematic, inflecting serious side effects in a percentage of the patients[3]. I think it's inevitable that we will knock this one down fairly quickly. In fact, I would go as far as saying that within fifteen years Alzheimer's will probably be a thing of the past."

"Wow, that's quite a conclusion. Referencing a failed clinical trial, out of a stack of failed clinical trials, you fast forwarded fifteen years and landed a cure. For almost two decades, regiments in the medical field have been intensely focused on making a dent in this most debilitating disease and the best they could do is delay advanced symptoms, but you think we will have a cure in fifteen years. Alzheimer's is a disease of the brain, and with all due respect to neuroscientists, we have not even scratched the surface as far as understanding how the brain functions."

"That is not true. Neuroscientists are way beyond the surface and new neuro-imaging practices are flourishing. Techniques like Positron Emission Tomography (PET), Single Photon Emission Computed Tomography (SPECT), and Magnetic Resonance Imaging (MRI) are going to unravel the mystery of the neuron jungle. Neuroscientists should be able to produce micro-detailed movies (not images) of the brain, in correlation with patients' real life experiences. I believe that the second decade of the twenty first century will be the decade of the

---

[3] Mathew Herper(March, 2002). "Elan Ends Alzheimer's Vaccine Trials". Forbes.com.

brain. We will probably discover in a year more about that black box than we have done in the past few decades. We will understand what makes us understand and abstract abstraction itself. We will be able to treat, nurture and even improve our brain abilities."

"Not to turn too philosophical on you, but I want to remind you that some in the scientific communities believe that, *abstractly*, this is an unsolvable problem, an NP-complete. The system cannot produce values bigger than itself. This is the conservation principle applied to nature, just like it applies to energy or thermodynamics. You cannot create energy. The human brain cannot create a more intelligent brain."

"Evolution has proven your proposition wrong. The human mental power has been steadily growing for hundreds of generations."

"You are missing the point. I am not saying that the human brain cannot be improved upon. All I am saying is that a human brain cannot create a more capable human brain. Nature, on the other hand, can. Improvements of that type will take centuries at slow incremental pace, not decades in the exponential fashion that you propose."

"We have been able to create machines that can perform better than humans for ages, from the Chinese abacus to Deep Blue, the undisputed world chess champion. In addition, the flood of advanced brain scanning techniques is going to bring us closer to our brains. Neuronetrix, an innovator in brain scanning systems, launched their COGNISION™ platform that can detect the negative cognitive effects of Alzheimer's using a technology called Event-Related Potentials (ERP's). This will help immensely in disease treatments. Assessment of the viability of a therapy or drug will become more direct, conclusive, and real-time. The Adler Institute for Advanced Imaging can now accurately pinpoint cancerous tumors in patients. This same technology may someday be able to illuminate the plaque in the brain associated with Alzheimer's disease and determine the impact of a drug to arrest the disease, according to Dr. Lee Adler[4], the medical director and founder of the Adler Institute. As the breakthroughs keep coming, we will eventually get to that junk which clouds our neurons and our judgment."

"Clearing the junk is probably the most you can do. Healing a chronic mental condition, however, is a different story. So although I will let you get away with junk flushing proposals, you are still far from curing Alzheimer's."

Herbert smiled. At least he got through to John, even if it was at low doses. He continued to further his cause.

"It gets better. Just like you, people have taken for granted that paralysis is an incurable misfortune. Once a nerve suffers extreme damage, there is no coming back, be it a broken neck, a broken back, or a serious head injury. Luckily for many, this is not true anymore. The wild card of this game is Embryonic Stem Cell (ESC) research, which promises rejuvenation techniques that can do more for anti-aging than any other approach. ESC deals with taking embryonic cells at the very early phase of fertilization, few days old, chemically altering them to incorporate the DNA of the target patient, and growing the resulting mix to morph into specific targeted cells (heart, liver, kidney, etc.). The end result: cells that not only match an adult

---

[4] "Advanced Medical Research May Lead to Breakthroughs in the Management of Alzheimer's". MedicExchange. www.medicexchange.com. 4/2009.

patient's signature (thus avoiding rejection), but are also young; embryonic young. It's like injecting a platoon of 10,000 young fighters into an army of old soldiers. Moreover, once the new vigorous cells are infused into the target tissue, not only they function correctly and blend in with the native cells, they also divide and introduce fresh new generations of young cells into the system. In theory, this seems very promising. In practice, results coming in have matched the hype. When was the last time that you heard of someone parallelized, from the hips down, regaining control over his bowl movement and walking the streets?"

I was sincerely taken aback. I did not know that we could do things like that, without voodoo and black magic that is.

"Are you serious? Are you saying that paralyzed patients can be cured?"

Nodding affirmatively, Herbert replied.

"Yes. What was squarely placed on the miracles shelf throughout history has happened a few times last year. Although the US has not been active on this front because of a 2001 executive ban instated by President George W. Bush, other countries have moved forward with the field and India is close to having such procedures routine. Fortunately, president Obama has acted fast on removing the restriction and we should expect the field to bloom very fast. You see, without the US engagement we can't expect great results quickly. The bulk of the medical research of the world comes from the US and a ban like the one Bush put on ESC has crippled the field. With Obama's open minded semi-liberal approach to the subject, I expect great strands for ESC in the coming years. Unlike other disciplines, this one not only promises great results, but aims for curing diseases that are grossly debilitating. Moreover, patients incapacitated with such diseases are desperately eager for cures. A neck-down parallelized person will probably be willing to give away half of his annual income ten years in a row for a chance to walk. A blind man will probably give up all his possessions for a chance to see, and a kidney-failed blood recycler will probably give everything he has for a chance to live without being hooked to a machine. This is what ESC promises in the near future, and that is why I am very bullish on its theme. Once people get the taste of what it can do and see cures in their fathers, sisters, uncles and neighbors, it will be impossible to stop that train. And the more this train moves, the faster it will go. As soon as we cure the Christopher Reeves of the world, we will be at a stage where growing healthy specialized cells for defective organs becomes routine. Every time an organ suffers in your body, you'll be able to give it a boost of youthfulness that can replenish it to its pristine state. Eventually, we won't use ESC because we medically need it, we will use it because it will make our lives healthier, more energetic, and more youthful. It will become a dietary supplement. This infinite supply of new cells that we will be able to summon whenever we want, will become a custom-made fountain of youth for the masses."

At this point, Herbert got my attention. By itself, finding treatments to tough diseases was an interesting subject, but creating an organ replacement process is something of a dream for so many, sick or not. It was borderline fantasy.

"Off the shelf body parts. Are we going to have them in sizes? Tin Man would have had a blast. Toto too. Even Hayman would have had a good brain; this way maybe he could recite the right formula instead of that triangular farce he did in the movie."

Then I turned to John and asked him.

"Did you know that they had a blunder in that movie?"

John sarcastically and condescendingly replied in a Johnny Carson style.

"I did not know that."

And in a triumphant voice I explained.

"After Ozzie grants Hayman his brain, he blurts out a statement that is supposed to be smart: *The sum of the square roots of any two sides of an isosceles triangle is equal to the square root of the remaining side*. Pythagoras would probably turn in his grave if he finds out that this is what we have been teaching kids."

"Ozzie? Hayman? Have you been smoking those cigars since you were five years old? His name was Scarecrow, as in those that you see in corn fields."

Herbert discounted John's comment and showered me with a proud-of-you smile and I sank back in the chair basking in my proud-moment. Then he went for the final stretch.

"And the last culprit on the list: Cancer. It is a type of cell mutation that causes cells to multiply without bounds. All what cancerous cells do is encroach on the space of neighboring cells, by growing fast and furious. That encroachment causes normal cells to lose grounds to malignant ones, making tissues and organs lose valuable healthy cells necessary for their function. With time, cancer cells theoretically should take over the body, but in reality the system breaks down long before that. However, advancements in cancer treatments have been tremendous. Fifty years ago, it was impossible to survive the disease. During the last three decades, scores of organ-specific treatments have emerged that improved survivability. ACS, the American Cancer Society, had set a goal of 50% reduction in cancer mortality by 2015."

John put his hand up to stop Herbert from going further.

"That was in 1996, and now it seems to be more elusive than ever. In fact, according to a report by members of the ACS's Ends Committee on Incidence and Mortality[5], it could take until the year 2040 or longer before we get to that goal."

"With the micro-level treatments being proposed or trialed, I believe we are still lined up for that goal. As bad as the disease is, it's the least of my worries when I think about combating aging. For one thing, I think the demise of cancer will be in early detection. I suspect that within two decades we will be able to detect most, if not all, cancers when they are but a few cells in size, and at that size, eradicating the harmful cells is not difficult. I suspect that it will become routine for all of us to take detect-and-destroy pills daily -- weekly or even monthly -- that will suppress the disease before it has the chance to flourish. In addition, with human genome sequencing becoming personal before the end of this decade, propensity to have the disease will be easily detectable by a simple genome test, and those in

---

[5] "2015 Cancer Mortality Goal Not On Track". CA Cancer Journal for Clinicians. 2006.

the highly-likely category can go through frequent rigorous testing to catch tumors at their early stages."

"Dear Herbert, you of all people must know that the aging-cure problem is much more than remedying a few diseases. Every time you close a hole in the system, others will open. Biology is very complex and theory is not enough. You'll have to see results in life extensions before you can claim success, and results of this type take a long time. Your trials will have to last fifty years or more before you can claim success. Life extension is a very hard subject and I doubt that we will be able to produce definitive outcomes anytime soon."

"Agreed, but keep in mind the following facts. First, the major killers of the world are very much known. Knock them down and you'll have a chance. Cancer, cardiovascular disease, and Alzheimer's disease can get you more than halfway through. Second, results have been seen in the labs with rats. Experimental results on their short life can be extrapolated to humans. Third, I believe that the first successes are going to be seen in the elderly. You don't have to wait for a fifty years old person to age before you make conclusions. You can start seeing results quickly in the ninety years old crowd. When you see them jogging down the road on their daily workout, you can start making useful assumptions. We are living in a different age now. What took hundreds of years to finish is accomplishable in decades, if not years. The next decade is going to have a molecular revelation, where scientists routinely poke the atomic level and read the script of life. This script has been tucked away in the nano-world for billions of years. Now the book is open and the records are readable. What, when and why will be answered and we are going to have a festival of theories that explain Biology's innermost mysteries. Just wait and see."

"And then what? Cures for all diseases? 175 years old grandpa will date nineteen years old cheerleader? What will happen to our society?"

"This is not the first change that we go through, nor will it be the last. Our ideologies have changed so much and so fast to make change the only constant in our lives (excuse the cliché). Less than fifty years ago, a woman's place was at home manufacturing children, and in the kitchen, feeding them. Now, half our society is run by women, and according to my wife, the better half. Early in the twentieth century, most people lived all their lives within a hundred square miles circle; now a 1,000 miles daily commute is not unheard off. If we can change from a segregated society, to one that elects an African American president, we should be able to handle grandpa dating Barbie."

"But how will we find resources for all these people? Humans on earth are like a punctured pot. Water trickles in from the faucet, and drips down from the crack at the bottom. You plug that crack and the pot will overspill. Three new lives come to the world every second. That's 100 million more mouths to feed annually. Compound that over the years and see what happens. Zero-down the death rate and you will have the already crowded Earth reaching a boiling point, and the fight for resources will undermine all the work you propose."

"If you shutdown death, and assume the current birth rate, human population will double in 35 years. I can live with that for many reasons. First, we won't make a dent in death rates anytime soon. Second, the birth rate is apt to go down by 50% as Asia and Africa numbers join the developed world. Last, *and most*, by the time our technological prowess enables longevity of the type we are talking about, that same prowess should enable other frontiers including growth in productivity, resource utilization and colonization. We should be able to get more, much more, from the base we have."

"Get more how? We don't even have enough oil. Prices have rocketed up to $150 a barrel in 2008. If shortages continue, we'll probably move back to bicycles. Double the number of people, and double the demand per person (which will happen by the time you double the number of people) and you have four times the pressure on resources. We are already launching wars over water, diamond and oil fields. According to the Washington Post, *the eight-year conflict that has shattered Sierra Leone and brutalized its five million people has been fueled by foreigners' hunger for diamonds. Rival mining companies, security firms and mercenaries have poured weapons, trainers and fighters into that country*[6]. The Global Policy Forum declared in 2009[7] that *More than fifty countries on five continents might soon be caught up in water disputes unless they move quickly to establish agreements on how to share reservoirs, rivers, and underground water aquifers*. And I don't have to tell you about the kind of wars we have had on oil fields."

"That is true, but this is going to be a race between optimal output and demand. Our inefficiencies are, by every measure, high. Just as an example, plants barely utilize 2% of the sunlight in the photosynthesis process. If we can somehow harness the abundant energy around us and channel it through optimum pathways, we might have all the resources we need at our

---

[6] James Rupert. "Diamond Hunters Fuel Africa's Brutal Wars". Washington Post. 10/96.

[7] "Water In conflict". The Global Policy Forum. http://www.globalpolicy.org/security-council/dark-side-of-natural-resources/water-in-conflict.html

fingertips. After all, eating food is a chemical process whose sole purpose is to create energy. In addition, our needs might become more contained in the future. The compulsion to have a car and go shopping might disappear altogether. In fact retail stores might become a thing of the past. True virtual reality concepts might become reality. Star Trek holodeck might become another room in your house, like the kitchen or living room. And when that happens, your need for leaving your house will vanishingly diminish. To top it all, we might be able to explore and utilize space on earth that we have never considered. We might start living underwater, underground, in the sky or even in outer space. The point is: we have much more space than we need or can use. We just have to learn how to use it."

"Herbert, you have to draw your reality lines and learn not to step outside them. Holodeck is a notion that will never happen. To submerge that deeply into reality is a concept that is better left to the imagination domains and the Hollywood writers. Same thing applies to living in space. As hard as we will try, we are not going to adapt; the tough problems will be impossible to solve. You will never be able to create gravity in space. You will never be able to have a human body survive zero-force environment, unless we live there for millions of years and adapt evolutionarily, maybe by growing long arms and slimy sticky material to hang on to ceilings and walls."

Herbert withdrew a little, like some chess player who still has a good game going, but is just about to lose a rook.

"It might be a while before we create settlements in outer space, but we have plenty of space right here on Earth. The Seasteading institute has already drawn plans to construct a floating city in the Pacific Ocean[8]. Such a city will not be any different from the ones on land. And remember, oceans cover two thirds of Earth. Also don't discount high risers. Burj Dubai is almost one kilometer high, Saudi Arabia is planning to build the one mile high tower in Jeddah, and we are just warming up. If the 13,000 feet X-Seed 4000[9] structure ever gets built in Tokyo, it is going to be a marvel of civil engineering. The eighth wonder of the world. In fact calling it a building is probably a misnomer. It's the artificial version of Mount Everest. With a six square kilometer base, it could potentially be the residence place of one million inhabitants."

"Here we go again. A building like this will require serious controls over pressure gradations and temperature fluctuations. This is going to be a nightmare in the making. And I won't even go into cost effectiveness. How much will such a mountain cost to build?"

"Probably somewhere between $500 and $900 billion, but that is not going to be more than it will cost to build homes, roads and infrastructures for one million residents. More importantly, the cost of operating it will be much less than the cost needed to operate a city with a population of one million."

By now, the cigar was one inch long and the heat of the flame was smoldering my lips. I plucked it out and sank it in the soda glass. John took this as his cue; he got on his feet and declared the meeting adjourned, or something along these lines. Actually his exact words were *As interesting as this discussion has been,*

---

[8] Jason Mick(Blog). "Engineers Planning Waterworld-esque Floating City at Sea". http://www.dailytech.com/Engineers+Planning+Waterworldesque+Floating+City+at+Sea/article14539.htm. 3/2009.

[9] Dorian Davis. "X-Seed Inspires Tall Tales". Architectural Record. 9/2007.

it needs to be parked on the to-be-continued shelf. Then he grumbled something about some meeting he had in LA and some flight he had to catch at 8:00 AM.

I sat down in the chair recollecting the discussion that passed, mostly with a great sense of optimism. Except for the last part, it was all good news no matter how you folded it. This Aubrey character must have had some serious issues with death, but then again, who doesn't. I mean to be given the option to live a couple of extra centuries would be marvelous, by all accounts, no matter what idealists or philosophers have to say about the subject. Deep down, everyone is petrified of that hole in the ground that will engulf each and everyone someday -- assuming you don't get cremated or your plane does not crash in the swamps of Florida where, in that case, your fear will be of the five or six alligators that will be first on the scene. Seriously though, to have superb medical tools that can create miracle cures is a dream by all measures. I gave Herbert a good-job wink and floated my hand toward the remote when Herbert, again, did an Olympic-worth goal-keeper dive and beat me to it. He turned the TV to – what else-- the Discovery Channel where they were running a documentary about *the instinctive intrinsic reactions of the mountain rabbit to the hunting rituals of the bald eagle*, which, if you ask me, can be summarized in one word: *run*. The tone of the narrator was more soothing than my mom's when she used to tuck me in. Few seconds later, I was sound asleep dreaming of Alzheimer's inflected, mighty superhero dog, smoking a cigar, and asking his buddy: "Tell me again, is it the mailman or the milkman that we're after?"

5

THE FUTURE

Welcome to the IT age[1]. The age where bits, stored on small physical medium -- for example a thumb drive -- are worth more than tangible objects, like a bedroom set. Information Technology has fused into our lives and become an intertwined part of everything we do. It's is not a tool anymore; it's at the center of almost every subject we tackle, from the simple, like writing an email to a friend, to the complicated like fabricating semiconductor wafers for the next-generation Intel microprocessor. While writing this paragraph, I had a laptop nestled in my lap, a blackberry on my right, an iPod in the drawer of the table next to the couch, four remote controls on my left, an environment control console on the table, an alarm system flashing green on the ceiling, a flat screen TV mounted on the wall, multiple digital video recording boxes on the table below, and an audio surround system spread around the room.

---

[1] IT is the acronym for Information Technology, and is not to be mistaken with the closely related acronym IT for Indian Takeover

The Future

Sometimes I wonder how we managed to do without the things we have now. How did we manage to meet a friend in the mall without a cell phone? How did we manage to construct a $10 billion, 50,000 employees program without email? How did we ever manage to write a book on a typewriter? I am sure our kids are going to have a much bigger how-did-we-ever list.

The whole IT trek started with a small device called the transistor; a tiny three legged creature that managed to crawl into our lives and embed itself all around us. Actually, it was not that tiny at the beginning. The first transistor was invented by a trio in Bell Labs, William Shockley, John Bardeen, and Walter Brattain, in 1947. It was the size of a palm. It went through a terrific journey over the decades that concluded with our ability to subdue ten billion transistors in that same palm.

So what is this transistor and why is it so important? The transistor has two separate functions: amplification and switching. Its structure is simple: a device that has three legs: one leg acts as controller for the flow of current between the two other legs. As an engineer in need of amplification, you connect the to-be-amplified party to the middle leg, the controller (the engineering name is base). Once the base is powered, the transistor starts conducting, allowing one leg, the emitter, to push its contents through the now-open valve into the other leg, the collector; and there you have it; a small amount of input opened the valve (made the semiconductor conduct) and let the stream flow between the emitter and the collector, thus amplifying the signal. Of course, when the base is not powered, the valve stays closed and the flow is disallowed. That same process allows the transistor to act as a switching element, and switching is the foundation of computer technology.

The birth of the transistor marked the beginning of the electronics revolution and an inflection point on the time chart of *"utilities effect on humanity"*. Building on the transistor concept, Fairchild Semiconductors developed the first integrated circuit in 1961, Gordon Moore put down his commandment stating that transistor density will double every two years in 1965, and Robert Noyce and Gordon Moore founded Intel in 1968, as if they did it just to make sure that Moore's predictions stay the course! Since then, and over a period of five decades, atomic level understanding and manipulation of material have flooded the world with breakthroughs that made 32 nanometer integrated circuits possible. Nowadays, our lifestyles are addictively entangled with IT which is reshaping every field, from the very core, such as computer science and integrated circuits, to the unlikely fringes of psychology and English literature. There is no doubt that our future and IT have become intertwined. What gets cooked in the technology kitchen has blatant consequences on the way we live. At the very least, fifty percent of what our kids do every day is different from what we did when we were their age. For example, they have moved on from the broadcast rituals of TV and radio, to the in-demand style of the Internet, podcasts, and interactive digital TV. They get so submerged in their portable game stations that you cannot tell that they are in the house (not that I don't enjoy a delightfully peaceful time every now and then). As long as computational power keeps growing at exponential rates, IT will continue to drastically change our way of life at an ever faster rate.

"Did IT enrich or blight our lives?" is the theme of a great many debates. What is an indubitable truth, however, is that IT has turned our lives upside down. I don't even have to look too far

The Future

back or think hard to find concrete examples. In fact, last night I went through a dizzying, circular, circus-like experience that made me feel that my life has become one big IT loop. I stepped into the house after a long day of being completely inundated by programming chores, and instead of being warmly greeted with the traditional hug-and-kiss, I received a stern command to program the DVR. I could have done anything for her except program a device. That day, I was programmed-out. And when I asked why, of all days, she urgently needed to videotape shows she told me that the Blockbuster video store next door went out of business and she did not have the time to schedule a recording for the Desperate Housewives reruns; why? Because she had to online-plan our Disney vacation next month and was raging a price war on the e-commerce travel sites; why? Because we were concerned about the way my daughter in the past few months had completely plunged into the Wii, DS, PSP, Playstation and Xbox worlds and was grossly missing on life; why? Because my brother recently discovered eBay and was buying every electronics game console he can get his hands on for his favorite niece; why? Because his wife had to run overtime counseling sessions at the *Here-To-Listen* volunteer institute with one of her patients and practically had no time for him; why? Because her patient's husband had been recently laid off and was causing all sorts of problems for the family, financial ones being the least on the list; why? Because his company, Blockbuster, had to close a dozen locations to keep up with the glut of competition from online-only sites and telecommunications companies (and those have apparently abandoned the Telephony market and were running hastily after the entertainment industry. Why they still call themselves *tele*communications companies, I have no idea). So just to recap, **Blockbuster closed doors** so wife had to tape, because she had no time, because she was eBooking a vacation, because daughter was playing video games, because brother was hooked to eBay, because his wife was counseling a woman, because that woman's husband has been laid off, because **Blockbuster closed doors**. Elton John's full circle of life, kind of. Or maybe Bill Gates full circle of Technology. Whatever it is, it has IT written all over it.

So take it from me, IT is here to stay. No one argues that the trend is not indefinite. Every good thing (or bad thing, depending on your view) must come to an end, but until that happens, we still have the opportunity to remodel our lives and rid humanity of all suffering, pain and need. IT is power, and just like nuclear power, it can have great impact on our civilizations. Well contained and guided, it can help us do wonderful things. The alternative scenario is as ugly as the Nagasaki bomb. We have the chance to write history the way we like our grandchildren to read it. Although the first few pages do not look promising, I am still optimistic.

The Dialogue

The night was long and sleepless. My one year old had just finished taking her shots, all five of them simultaneously. What it was doing to her body and mind, I can't tell you, but I can vividly tell you what it did to the tranquility of our bedroom at three o'clock in the morning. After tossing and turning from one side of the bed to the other, trying every possible pause, pillow over the head, head under the blanket, pillow over the head under the blanket, I decided to call it the night at 7:00 AM. While tiptoeing through the little beds scattered in my bedroom -- my one year old crib, my five years old daughter's bed and the beds of their respective dolls – I stepped on one of those very small, yet very sharp Cinderella rings that were made to

The Future

penetrate, and it was all I could do not to scream at the top of my throat. I have no idea why they put all kinds of warnings for parents about safety precautions and risks of kids choking on small objects, when the only purpose I see for these little objects is to stab the belly of an adult foot inducing unbearable pain. If you ask me, this is how the underground spy community should torture their foes. Forget simulated drowning. Just put the interrogated spy at the end of one room, in hot weather, with a 2-liter water bottle, and separate the tortured from the much craved bathroom with Disney characters rings, McDonald happy meal toys, and a few Star Wars sabers. Within six hours you will have the locations of all double agents in Europe.

Anyway, I jumped groaningly on one foot, with the other one in my two hands, into the kitchen and headed straight to the coffee machine. I needed to wake up. Of course, we were out of coffee. We don't run out of coffee when late at night my wife asks me *"dear, would you like to have a cup of coffee?"* and I pause for a few seconds thinking it over, contemplating the lateness of the night and wondering if I really should have caffeine this close to bedtime. We run out of coffee after a long night like this one, when every cell in the body is aching for the refresher and the coffee pot seems to be a sight for the literally sore eyes.

I jumped into the shower for a nice hot one, put my clothes on in ninety two seconds and in no time was out in the car heading to the closest coffee shop. There was a Starbucks couple of miles down the road from my house and, although I am not much of a fan for their exotic coffee, two miles was all I could do without caffeine. I staggered into the coffee shop and made my way into the zombie line. Few minutes later, I was next. I looked at the menu and could not understand a thing. So I started with a question.

"Good morning. I am really tired and need a serious shot of caffeine in the head. What do you recommend?"

She looked at me for a few second, probably sizing me up and running a drink-match program in her head, and then she gave me her recommendation.

"Good morning sir. Try the double ristretto non-fat organic chocolate brownie frappucino, double blended, shaken not stirred."

Not understanding a word, I replied "alright, let's have it."

"What size sir?"

Glancing at the menu again, the first thing that stood out was tall. Tall sounded about right and tall was what I asked for. Ten minutes later, they called that complex drink name and I headed to the delivery table only to find a small drink waiting for me. I called the lady and inquired about the size of the cup and she explained to me that *tall means small*. The sentence begged for a retort, but I realized that without having that *little* drink, I won't be able to put up any arguments. I grabbed the cup and headed to a nearby table, when right in my view came John and Buzweil, apparently having their Saturday morning coffee get-together and, from the look of it, they were submerged in some serious debate. I approached their table and hello-ed every one. All I got back was a glance of acknowledgement and a nod, so I went to the next table, picked up a chair and tried to squeeze a seat with the gang. If you

The Future

know the Starbucks round yellowish tables, you'll understand what I mean by squeezed. There is barely enough room for one person to sit down and hug the table. What ends up happening is that everyone sits a yard away from the table, in order to make legroom, and occasionally reaches in to grab his cup off the table. After a few iterations of awkward reach-in, take-a-sip, and put-cup-back, everyone decides to keep his drink in his hand and the table becomes pointless. I tuned into the intense discussion and Buzweil was passionately making a point.

"IT is the most distinct change catalyst in the world, and its influence is far reaching. There isn't a field, science or art, that hasn't been hauled around by the IT storm. You can't find a professional or even a student, trekking through his career without being consumed in his laptop. And this is only the beginning; advances in microelectronics…"

And John apparently knowing where Buzweil was heading, interrupted.

"I beg you, please, do not throw that Moore's Law again in my face. I can't take it any Moore."

"Don't make fun of one of the most accurate trend-predicting statements of our times. The doubling of transistor packing efficiency every eighteen months created Very Large Scale Integration capabilities which in turn fueled the IT super-engine. Doubling is an exponential event that produces sharply rising curves. It has been running steady for more than four decades and is in the process of clearing the fifth with no hurdles."

"You'll have to admit though that its sustainability in the future is debatable, and even doubtful. There are more announcements about the death of Moore's Law than there has ever been since it was published in the sixties."

I couldn't understand what the big fuss was about. So we doubled the number of transistors that can be packed in my computer. Why is that so significant? I stopped Buzweil before he had the chance to answer John's question and squeezed in.

"Why are you guys so fiery about this? What's so earth-shattering about doubling transistor density every eighteen months?"

John looked at me as if he just realized that I was there and sarcastically said.

"Good morning pretty boy. I didn't notice that you sneaked in. Look, why don't you just enjoy your cup of tea in peace. By the way, how on earth did you ever manage to crawl out of bed before ten AM on a Saturday morning? I thought we were pretty safe from you today."

I gave him one of those show-your-teeth smiles; the ones that says *ha-ha, very funny wise guy* and with a touch of spite I said.

"For your information, this is not tea. It's a fancy caffeinated drink with a name longer than I could possibly remember, but it is not tea."

I knew that my comeback was pretty feeble, but it was all I could dish out on a Saturday, early in the morning, and after a long sleepless night filled with baby cries. I still wanted to know what the big deal was about cornering tons of transistors in a tiny space, so I turned to Buzz and asked.

The Future

"But seriously Buzz, why are you so hopelessly in love with this Moore dude and his law? What's so grand about it?"

And Buzz obliged.

"You see, although doubling is very hard, it's very powerful. Ten doublings give you a factor of 1024. Twenty doublings is roughly a factor of million and thirty doublings is a factor of billion. According to Moore's Law, your computer could have the power of all computers on the planet put together within thirty or forty doublings. With such powerful engines, there is no limit to what you can do."

Then Buzz turned to John and resumed.

"Intel, as of January 09, announced the arrival of the 32 nanometer integrated circuit. Few more realizations of the sort and we will have machines sitting on desktops that can execute faster than the fastest networked-supercomputer of today. Just imagine what will happen to the networked-supercomputer."

"Listen Buzz, I understand your excitement, but you have to stay pragmatic. We are about to hit limits that we can't overcome, and at that point, your Doubler will come to stand still. You'll be lucky to trickle a 10% increase every year. An atom's diameter is in the range of 1 angstrom, that's 10^{-10} meters or 0.1 nanometers. To sustain Moore's Law we'll need to get to 16 nanometers levels within two years, 8 nanometers within four years and 1 nanometers within ten years. Five doublings (or halve-ings in this case) and we'll be fiddling with atoms; at that scale no scientist can do any meaningful work for many reasons. First, objects that small will be hard to manipulate and therefore processes cannot be controlled. Second, at that size, Herbert's weird quantum physics takes center stage and things become fairly unpredictable. And since you were almost half asleep when Herbert and I were talking about quantum physics let me enlighten you. Quantum physics is very hard, very unpredictable and extremely incomprehensible. That is consensus. From my prospective, quantum physics is incomplete and is waiting for the right mind that will come and put every bizarre aspect of it in order. These bizarre aspects are the hinge-pins for the exponential improvements that scientists are dreaming of. And even if we manage to dabble with the quantum genie, and go down to the atomic level, which is extremely unlikely, we will run out of doubling space very quickly."

"That is true. At the smallest of scales, at the atomic nucleus level, quantum physics governs the laws of behavior and that's why quantum computing could be the next big frontier. Quantum computing promises to do to the integrated circuit what the transistor did to electronics. When we solve the remaining Qubits problems, we are going to see a shuttle-launch-like climb in the curve of electronics performance."

John shook his head and looked at me as if saying *What's wrong with this dude*, or some variation of that, given that John abhors the word dude. I don't know why; I mean dude is the most versatile word in the English language and I use it in like every other sentence; but that's John. He is too formal a *dude* to be heard uttering words of that low rank. He then turned to Buzweil and coarsely said.

"Are you half asleep again? I take it back. You are definitely sound asleep. Didn't I just finish reciting my quantum physics

The Future

scorn-report? Quantum computing is not going to happen before we fix quantum physics. The outrageous quantum mechanics theories that the physics mavericks are proposing are fanciful fictions. Face it Buzz, Moore's Law is dying and speculators like you are trying to clutch to far-fetched concepts to convince the world that it is not. Len Jelink, a director and chief analysts at Isupplie, a market research company, predicts that Moore's Law will die by 2014 because *usable limit for semiconductor process technology will be reached when chip process geometries shrink to be smaller than 20 nanometers*[2]."

"And you are quoting a market research company on the future of computing?"

"How about this? Carl Anderson, an IBM fellow at the Systems and Technology Group, states that Moore's Law is in its golden years. Do you know what an IBM fellow is? It's the most prestigious scientist title in the most prestigious electronics research institute in the world. Is that worthy enough for you? Anderson asserts that, just like railroad and aircraft industries, semiconductor devices will hit their exponential growth limit. According to a report published in EE Times, Anderson states that *a generation or two of continued exponential growth will likely continue only for leading-edge chips such as multicore microprocessors, but more designers are finding that everyday applications do not require the latest physical designs*." Moore's Law is not going to sustain indefinitely. Fast forward fifty years and tell me how you are going to have semiconductor devices packing-capacity improve by a factor of 2^{33}. That's a factor of 8 billion. That's 40 trillion transistors per square inch."

---

[2] Michael Feldman. "Moore's Law to die at 18 nm, analysts predict". http://www.hpcwire.com/blogs/The-End-of-Moores-Law-in-Five-Years-48287682.html. June 2009.

"The spirit of Moore's Law is more about the power that could be delivered by devices than it is about transistor density. Transistor packing trend will probably carry us through 2020, but by then we will start riding other waves. But before we skew to a different track, tell me something, why do you think it is hard to work with quantum physics?"

"Because we don't have the skill to see how things work down below. The well is so deep that for all practical purposes you could think of it as bottomless. We just don't have enough intellectual strength to watch the subatomic particle game. We are almost blind when it comes to dealing with the nucleus of the atom. Billions of dollars are being spent on building gigantic particle accelerators that can barely give us a glimpse at the quark-level show."

"But we will. Every trend in the world is up and the factors that help this exponential trend sustain are plenty. For example, consider the number of people in the scientist community. Fifty years ago, researchers contributing to the fields of science were scarce; the science domain was limited to western civilizations and even there, the contributing institutions were small. Nowadays, scientists are popping up in the farthest reaches of Earth and in abundance. India was nowhere on the map of scientific conferences twenty years ago. Last year, you could not keep up with the announcements of major conferences happening across that country. The world has become flat and the output of the scientific community has grown exponentially, creating a web of research connectivity that meshes the globe. In addition, this overwhelming complexity creates fertile soil for spawning artificial intelligence. Whatever we create deterministically will always be limited by our abilities and will always be sub-human. However, the intelligence that will grow

The Future

out of the complex IT jungle that we create will be overpowering. AI will be the helping hands that will enable us to solve the problems of the infinitesimally small world. When we managed to understand the atomic structures, we were able to create powerful machines that fueled the world. When we go one level deeper into the sub-nucleus level, we might discover powers that we can only dream about. That's how we are going to keep the Doubler thriving and that's how we are going to launch quantum computing."

I took a big sip from my coffee cup while listening in the background. That quantum computing stuff seemed interesting, but I had no idea why Buzweil thought it's a game changer. It was too early in the morning to ask and I knew that by asking, I will risk receiving a convoluted physics lecture that will require much more than that puny caffeinated drink that was handed to me by that Starbucks coffee-drinks connoisseur, but I took a chance.

"Buzz baby, why do you think that quantum computers are much more powerful than classical ones?"

"It's simple; counterintuitive but simple. Assume that you built on a silicon wafer a ten-bit register; a placeholder that can store ten digital bits. Although any of the 2^{10} –1024 – possible permutations can be stored on that register, only one will be set at any point in time. To be more specific, any time you check the register, you would find one value imprinted on it, e.g. 1001100110."

"Alright, straightforward. So?"

And John sarcastically took over from Buzz.

"You are not going to like what's coming. Now enter the quantum Land of Oz. With quantum computing, ten quantum bits, or qubits, can also store 2^{10} –1024 – permutations, but they can store them all at the same time."

When John finished his last sentence, I did not know what to think.

"At the same time?"

And Buzweil replied.

"Yes, at the same time. That's called quantum superposition. So, to be more specific again, at any point in time the register can have the values: 1001100110, 1000100000, 1111111110, 0001000000, and every other possible value on the 1024 permutations list."

I wrestled with the concept for a few seconds, and tried to view it from every imaginable angle, but nothing clicked. I could not connect the dots. In fact, there were no dots to connect. This is physically impossible. How can you store 1024 different values on the same bit location? I then turned to Buzz and told him my candid, well thought opinion on the subject.

"That's crap."

And Buzz did not like my attitude.

"That is physics that was created by the most intelligent minds of the twentieth century."

The Future

"Buzz, dude, my friend, how can you store two, let alone 1024, values in the same location?"

"This is how things work down there at the quantum level. The nano-universe does not align with our common sense."

That's it? That's all the explanation I get. We will accept an utterly unacceptable enigma because *the nano-universe does not align with our common sense*. This seemed to be a rerun of the discussion we had last week. I was OK with some unconventional physicists proposing outlandish theories and models to describe the atomic-level playgrounds, but I did not expect Buzz or Herb to believe that consequences of such theories could prompt such far-fetched implementations in the real world. Before I had the chance to shower Buzz with a few *are-you-out-of-your-mind* comments, John beat me with his sarcastic question.

"So you expect to build things to service us in our common sense world using tools that do not make sense. Does that make sense to you?"

The discussion stopped there for a few seconds before Buzz, feeling that he was down on his points, threw another chip on the table.

"Even if quantum computing does not work, we have other weapons in our physics arsenal."

And John, as if eager to get off the quantum computing subject asked Buzz what else did his *arsenal* have and Buzz dove into his spintronics monologue.

"I'll start with a good prospective. Special materials, called topological insulators, allow electrons to move on its surface without energy loss. If you use the electron's spin -- a property of the electron -- you can transmit information and potentially build electronic devices, like transistors. It's called spintronics and devices built on this concept can lead to great advances in microelectronics. With the spin acting as a bit carrier, we can develop devices that are superior to the ones we have. For example, information that we store on charged particles disappear as soon as the power is turned off. Spin, however, is non-volatile; even if unplugged, it still keeps its value. This could have great applications in starting devices instantly. Also spintronics devices do not consume much power, which means that large nano-electronic instruments can be built without the risk of overheating."

"Fine; and you'll have a low power computer that starts instantly. Big deal. Is this all you have?"

"You are missing the point completely. Electron, the fundamentally small subatomic particle, will act as a switching element. That will keep Moore's law breathing a few more decades. In addition spintroincs is not the only game in town; a long line of microelectronics companies are working diligently on launching new efforts that will keep the growth in performance going at the same level it has for the past five decades. For example, Intel employs more than 1,000 researchers focused on bringing to reality the next generation of technologies."

Buzz was silent for a few seconds before he drove into yet another hotly debatable concept.

The Future

"And if these approaches fall short, nanotechnology should make up for the slack and keep the technology locomotive running at the required speed."

Without even knowing what nanotechnology is, I knew that it won't rhyme with John, and he did not fail to be consistent.

"Nanotechnology is the scariest concept contrived since quantum theory unleashed the power of the atom. It's a fundamental truth that the further we protrude into the inner cavities of nature's infinitesimally small provinces, the more powerful things get. That is how nuclear power was created."

A question that has been lurking in the back of my mind for a while now popped to the front as soon as John announced his detestation of nuclear power.

"You know, I just don't understand this. How can there be so much power in something as small as the minute nucleus of an atom?"

"Before we discovered the *strong force*, which binds neutrons and protons together in the dense nucleus, there were two known fundamentals forces: *gravity* and *electromagnetism*. We did not know about the strong nuclear force because it acts on very small particles, protons and neutrons, inside a very small entity, the nucleus of an atom, across very short distances, 10^{-15} meters (that's 0.000,000,000,001 mm or, using what seems to have become a standard unit of measurement, that's one billionth the diameter of a human hair). Available only for protons and neutrons, these forces are too small to be felt, noticed, or even measured, but when they are applied to the conglomeration of atoms that can be lumped in visible mass (for example, 1 liter of water packs 10^{26} atoms), the force becomes the horrific nuclear power that we all dread."

Naturally, Buzz did not agree with the association that John was making between small and dangerous.

"But nanotechnology is all about making use of these forces in a structured manner while nuclear power was all about blindly blasting the whole to emit explosive forces. At the heart of nanotechnology is the process of constructively assembling armies in the nano-world to service our needs. It's about recruiting trillions of atoms and molecules to become an assembly line that manufactures goods. Nanotechnology is much less dangerous than nuclear power."

"Nanotechnology is more dangerous than nuclear power by leaps and bounds. There is no doubt in my mind that nanotechnology could have the potential to unleash amazing amounts of power, but there is also no doubt in my mind that the path to nanotechnology is very difficult, if not insurmountable. I just don't see a feasible and practical course that could lead to the creation of efficient, reliable, cost effective nanotechnology implementations."

"The success of nanotechnology will pretty much depend on the creation of an assembler, which is a device that can build products from the basic building units. Everything in the world is made of the 115 known elements. Theoretically you can build anything you want as long as you have enough raw materials in the form of elements[3]."

---

[3] For example, aspirin, the off-the-counter pain relieving medicine, has the chemical formula C9H8O4. Nine molecules of carbon, eight molecules of Hydrogen and four molecules of Oxygen is all you need, from basic material prospective, to assemble

The Future

"You will need to execute a chemical reaction, a carefully orchestrated process that leads to the transformation of one set of substances into another set, under certain environmental and catalytic conditions in order to build any substance. For an assembler to work, not only it will need to have vast access to raw material, but it needs to house all the blueprints of the needed chemical reactions that will lead to the manufacturing of the material needed."

That was convincing. I turned to Buzz and asked.

"So is this assembler a valid concept that is being pursued in the scientific circles, or something that you just concocted?"

"A molecular assembler was first introduced by Eric Drexler in his 1986 book *'Engines of Creation'*[4]. It's a precise machine capable of creating any object from the basic constituents of the universe, the elements. The concept is still in theoretical stages and has a long way before it gets to practical, on the ground, implementations."

"Let me translate, a molecular assembler is a fiction-like concept that builds on practically impossible credible theoretical ideas. Assembling an object, molecule by molecule, is an infinitely exhausting task that requires fantastic precision and super natural speeds, two almost contradictory terms. Assuming that we overcome all the technological problems associated with understanding and storing the fundamental

---

aspirin. Practically, however, the process of preparing a substance, like aspirin, involves a lab chemical reaction like the one listed below: salicylic acid ($C_7H_6O_3$) + acetic anhydride ($C_4H_6O_3$) → acetylsalicylic acid ($C_9H_8O_4$) + acetic acid ($C_2H_4O_2$)
Acetylsalicylic acid is the chemical name for aspirin.

[4] K. Eric Drexler. "Engines of Creation". Anchor Books. 1986.

structures of a machine, fantastic precision requires slowing down, and fast moving tasks imply serious lack of precision. Eighteen grams of water, which is 0.65 ounces, necessitates assembling three times Avogadro's Number in atoms (every molecule of water has two hydrogen and one oxygen atoms). That's $3 \times 6.02 *10^{23}$ atoms to assemble something as big as a baby mouthful. Now, if you have a machine that can assemble one trillion atoms per second, it will almost take two trillion seconds, or 63 thousand years to put together such relatively small amount of matter."

"You are right; assembling at the molecular level is extremely hard unless the process is self-running."

Skeptically, and almost as if he knows the kind of answer he is going to get, John asked.

"What does that mean? How do you propose that we create a self-running assembler?"

"Self-running assembly processes are the way of the world since the inception of life. The best example of an efficient, intelligent, and reliable self assembling machine is the human body. It starts with one cell that assembles itself into one hundred trillion cells in less than twenty years. Paul W.K Rothemund, an IBM researcher at the California Institute of technology is focused on the infusion of biologically controlled chemical processes to advance semiconductor fabrication technology through DNA origami. On October 2009 he reported progress that could lead to a major breakthrough in enhancing our ability to fabricate record-levels dense electronics on a chip through an economical, efficient, DNA enabled manufacturing

The Future

process[5]. Rothemund combined lithography with self assembly to enable existing semiconductor equipment to delve into the sub 22 nm manufacturing process. The DNA molecule was used as a support structure --scaffolding-- where carbon nano-tubes were deposited to assemble themselves. This biologically driven method of manufacturing can lead to cheaper fabrication processes, thus removing one of the biggest barriers of diving below 22 nm: cost. As we get closer to the atom diameter size (0.1 nm), the fabrication process is becoming prohibitively expensive and therefore impractical for mass production. Through DNA Origami, Rothemund hopes to make DNA molecules self-assemble, allowing them to act as staples that can modify structures and create geometries like triangles, circles and squares, and eventually, transistors. Rothemund demonstrated the power of his techniques by writing the letters "DNA", drawing an image of a double helix, and creating a nano-map of the western hemisphere in which one nanometer represents 200 kilometers. Although practical applications of the technology will probably take years to mature, this step is a clear proof-of-concept that biologically enabled self-assembly is a powerful paradigm that can change the map of electronics fabrication in the future. In essence, this is the first stage of a Drexler assembler."

At that point I heard enough. I really did not want to turn this into a biological microelectronics DNA origami chemistry lecture.

---

[5] Veronique Greenwood. "With a few strands of nucleic acids and some ingenious programming, DNA origami is remaking nanotechnology, from drug delivery to chip design". Seed Magazine. www.seedmagazine.com. October, 2009.

"Spare me the details, please. Just tell me, where do you think all of this is heading? What do you think these technologies will do to the bottom line?"

"I won't give you my opinion. I will share the views of one of the most successful technology trends predictor of our time. Ray Kurzweil is an inventor and a futurist who has been called the successor to Thomas Edisson. He was described by Forbes as *The ultimate thinking machine* and Bill Gates says that *he is the best person I know at predicting the future of artificial intelligence*. In his book *The Singularity Is Near*[6], he outlines how our future is going to look like. By 2011 almost each one of us will be completely dependent on computers which will essentially become extensions of us."

And John, who seemed eager to argue even before he heard the sentence, jumped on Buzz immediately.

"I don't think that this will happen any time soon. Even if such technologies start to appear, people who already invested in a computer are likely to stick to their investments a couple of years down the road."

"That's not the point. Your desktop will stay snuggled in your office, but it will be overpowered by the glut of devices that will surround you wherever you go. Increasingly, you will have computers on your wrist, around your neck, in your briefcase, inside the TV, around the kitchen, over the gate, in your car and almost everywhere you go. Smart Cities, Smart Villages, Smart Buildings and Smart Homes have become the rage. They are new spinoff disciplines; the offspring of the marriage between

---

[6] Ray Kurzweil. "The Singularity is Near". Penguin Books. NY. 2005.

The Future

construction and computer technology. All new structures are being wired to accommodate IP (Internet Protocol) technology, and IP has become the nervous system of our society. It fuses in the background of the average home and acts as a medium for applications that allow residents to connect to the hive. In the near future, not only we will intelligently provide services to the occupant, but we will also save him money, creating viable business models that will help propel the Smart Environment concept to the forefront. For example, your water heater could be adjusted to turn on one hour before you wake up and one hour before you come home; your shades, AC and lights will be controlled centrally from a console that fits in your hand, just like the TV remote control. Security cameras will be viewable not only on your TV, but also on the Internet, and by default on your cell phone, in case you want to check on your house when you are at the office or when you are on vacation. Your garage door will open as soon as you come close to the garage; your environment will be set in your house as soon as your car is parked in the garage; your doorbell turns the TV on and links it to the gate camera; and tons of other features that started to crawl into the average home. That's just home automation. Building Managements Systems, BMS, and Facility Management Systems, FMS, are completely different beasts. The process of collecting your power, water and central cooling usage for your apartment, aggregating the data and churning out a bill that automatically gets paid by your bank account requires no intervention from you (except probably for you to check the bill on your bank account at the end of the month and argue with your wife on why the AC was kept on during your vacation trip to the Bahamas). All of these are products readily available and many are already implanted in many new structures. You see, the sheer number of undetectable devices that will surround you at any point of time will overshadow your desktop. In addition, home automation will take on a different meaning. As robot technology keeps growing, it's going to become a crucial part of our life, especially on the housework front, one of the most time-consuming and distracting aspects of our lives. Automated maids at homes will have more impact on our society, per capita, than any other technological tool."

"Now what does this mean? You can call the Roomba vacuum cleaner an automatic house cleaning robot. An advanced version of the Roomba will be of some help, but a Robot that can do the work of a maid is not happening any time soon."

"Why not? Just like the Roomba, there could be a set of small specialized devices that can do specific things. A glass cleaning Roomba for example; a nifty sink that sorts dishes, places them into the dishwasher and washes them; an automated ironing device that takes clothes out of the drier, irons them, and folds them neatly."

"Ok, but as long as you and I agree that there is no intelligence here; just well designed machinery."

"You could call it dumb, but I bet you that housewives will call it a genius. It's the first step towards robotizing the world. A combination of such well-designed machinery and biologically-integrated instruments in our bodies and brains will push humanity to a different frontier. Nanobots infused in our bloodstreams will make us cyborgs, according to Kurzweil."

You could not keep John on low-volume mode anymore. The tone went up a couple of levels and the hand and body gestures matched the tone. It was getting a bit risky sitting in the middle, so I pushed my chair a few inches back, into what seemed to

The Future

have become the front seat to a heavyweight world boxing championship, while John started his mock session.

"No we enter Lala Land. Have you seen what it took for us to get anything approved through the FDA? Do you honestly expect that in ten years we will be able to put foreign nano-cells in the human body through a medical procedure that will in essence transform us into Cyborgs. Look at the kind of progress we made between 2001 and 2010. Have you seen any quantum shifts in the way we treat people medically? Why do you expect the gates to fly open in one decade? Because of Moore's Law?"

"Moore's Law will help, but there are other key factors. During the past ten years we have waged an all out technological war on medicine. We sequenced the human genetic blueprint and we are about to personalize it. It won't be long before we will have a databank of all patients' genomes, and the tools to correlate that data. This is the first time in the history of humanity that we see inside clearly, and at the most granular level. We are going to produce an amazing wealth of information in the coming few years and we are going to formulate a new medical science that makes everything we have done so far obsolete. Propelled by artificial intelligence, we will open a new page in the way we understand the world and live our lives."

"Now forget everything else and focus on this one. You take it for granted that AI will arrive. Of everything you are blabbering about, this is the pivotal point. We don't even know if AI is attainable. The arguments for everything that follows will rely on having machines do the work that we obviously are unable to do."

And almost frustrated, Buzz looked at John for a few seconds and then said.

"Why is it so unacceptable to you that we will land AI within the next twenty years?"

"For fifty years, we could not make a small dent in that field, how do you really expect that suddenly, within a couple of decades, it will all be clear."

And I stepped in. I did not want to have the same discussion on AI that we had a few days earlier, but I was getting excited. Could this be for real? There has to be something to it if this well-decorated brain dude Kurzweil thinks it's inevitable.

"John, I think we already discussed this to death last week and I really don't want to have a regurgitation session. Let's see where all of this is going to lead. I really want to hear how far this Kurzweil brain-dude will go."

Then I turned to Buzweil and said.

"Buzz, do continue with the predictions and don't hold back anything."

"Once humanity and machinery integrate, activities like mind uploading will be possible, opening a whole new frontier for the advancement of science."

"Mind uploading, right, sure, by all means. Are you out of your mind? Do you know how hard it is to come close to our brain? There is a reason why *Brain Surgery* is cited every time someone needs to quote complexity. It's the riskiest, most delicate type

The Future

of medical operation that surgeons perform, and you want to tap into the brain and upload data. Get real, will you?"

"Brain scanning techniques have advanced so much in the past few years that even Moore's Law is considered too slow for it. In addition, and for your information, scientists now know how to build bridges in the brain to fix neurons and restore brain functions. Rutledge Ellis-Behnke and Gerald Schneider of the M.I.T. Department of Brain and Cognitive Sciences have performed marvelous research on hamsters to help re-establish bridges between disconnected brain cells. Using nanotechnology, the bridges were allowed to assemble themselves from protein fragments called peptides after being injected into the injured area. The results were stupendous. According to an article in Science Central[7], *hamsters with severed optic nerves, axons -- the long arms of brain cells -- grew across the bridges and re-established connections, restoring partial sight in the hamsters.* Remember, this is now, and we still have twenty years to get there."

That pretty much shut John up and got my interest level in the discussion to swell to a mountain size.

"Go ahead Buzz, what's next."

"Well, according to Kurzweil, Human Body 3.0 will arrive soon afterwards, that will in essence create a new species made of the merger of man and machine."

---

[7] "Brain Healing Bridges". Science Central. April 2006

"I think I heard enough. Are you listening to yourself? Human Body 3.0? You think we are going to be able to upgrade our parts?"

And I kind of agreed with John, but Buzz stood his grounds and defended Kurzweil like an army soldier defends his hometown, and, to be frank, he did a good job. Good enough to make me get even more excited about the whole prospect.

"Bionic implants and human enhancing devices are commonplace now. In fact, as early as 2002, the National Science Foundation issued a report on *Converging Technologies for Improving Human Performance*[8] that discusses nanotechnology's contributions to the advancement of the medical field in the next two decades. The report is not limited to prosthetic limps or simple implants. It dives deep into the prospects of integrating technology with the internal functions of the body. So don't tell me it's too much. I suspect it won't be long before losing an arm or a leg might not be as big a deal as it used to be."

I looked at John trying to read on his face why he did not like what he is hearing. So far, all these predictions seemed worthy of celebration. Before I had the chance to inquire about John's attitude, Buzz continued with his recital and pretty much answered my question.

"By 2045, the singularity will hit, at which point, machines integrated with humans will govern the world and set the path of the future."

---

[8] Mihail C. Roco, William Sims Bainbridge. "Converging Technologies for Improving Human Performance". National Science Foundation. June 2002.

The Future

And my emotions changed gears from excited to terrified. I saw clearly why John did not like the subject. I turned to Buzz and, almost scornfully, said.

"This is scary. You just walked me through a plausible scenario that ends up with machines having the upper hand and pretty much ruling the world? I know we had this discussion earlier last week, but let me ask again: What will happen to us if this becomes a reality?"

And John emphatically answered.

"I'll tell you what will happen to us. We will be squashed like cockroaches. We will be incredibly weak, disgustingly stupid and hopelessly outnumbered. Like I said, this will only happen if we attain AI, but here is some news for you: We won't. If we do, I will be the first one rallying in the streets carrying a billboard-size sign that reads *Unplug the iron monster.*"

"I am more hopeful than you guys. I think machines will have better things to do than destroy us, like what happens Post-2045. According to Kurzweil, at that point, machines will evolve at unparallel speeds and will start eating up outer space in an effort to create a cross-universe entity that wakes up the cosmos and controls time-space."

"Buzz, let's come back to Earth. The major debate is going to be about the chances of conquering a few stumbling blocks like Robotics and AI. Both are very hard. In fact, AI might prove to be beyond reach."

I did not understand why these two concepts, among all the far-fetched weird prospects that Buzz catalogued, were decisive.

"Excuse my ignorance, but why are AI and Robotics the central themes here?"

Buzz looked at John, as if saying *would you like to explain or should I* and John rolled his eyes and said.

"Robotics is the first step. It will relieve humanity of the burden of the mundane. Once we have robots that are capable of running routine work, life will change. I'll give you an example, imagine having robots that drive cars flawlessly, clean and tidy houses entirely, scrub the roads, build buildings, manufacture products, sell services, work as cashiers, and completely master every other specialized, simple-minded errand. You will have orderly cities, houses built in a day, cars manufactured in hours, traffic gridlocks eliminated, and so on. Every boring, routine, time-wasting chore that humans do will be attended to. You will have a society freed up to perform higher thinking and focused on advancing the state of the art. You will get closer to AI."

"Alright, but that will not *wake up the universe*, will it?"

And Buzz, who was having trouble being on the bench when his favorite topic was being deliberated, burst into the scene.

"AI will. AI is intelligence at the human level."

I still could not figure out how they got from AI to spaceships and living universes.

"You do all of this to get machines as intelligent as we are. How would that help you with your fantastic inter-dimensional cosmic-dismembering voyage? These machines won't be able to do more than what we can, will they?"

"Yes they will. First, a machine that attains our intelligence will also have impeccable gigantic memory, fantastic ability to do computations, and sponge-like aptitude to absorb information. Machines won't tire, won't stop for a break, won't need vacation, won't get sick, and won't have to sleep. They will run 24x7 focused on one and only one thing: Advancing. Imagine plugging this machine into the Internet; within few hours, or maybe days, it will attain all the knowledge of humanity. It will have the skill of the most astute scientists, put together in one brain. That machine will quickly become more intelligent than all humanity put together. Put a few scientists from different disciplines in one room and they will come back, probably within one day, with intriguing concepts. Synergy will produce miracles. Now imagine all the scientists in the world with all their brain power concentrated in one shell. You can't even conceive what will come after that. In addition, the first thing such a machine will do is create a system more intelligent than itself. It can because it is more intelligent than its creator. Keep this going for a while –a powerful intelligent machine diligently working on creating a more powerful machine -- and the next thing you know, you are in the presence of an entity whose intelligence we can't even comprehend. How it will wake up the universe, I can't tell you, but I can tell you it will. It will take the world by storm and what it will produce is pretty much an unknown. To predict the products that it will spawn will be like having a rural man from Byzantine times contemplating the future of *QMOS Plausibility for Deep Submicron ULSI Design Paradigms*; it is simply outside our limited mental capabilities."

At that instant, I did not know what to think. I tried to take a sip from my tiny cup, but it was empty. I headed to the counter again, took another survey of the menu and ordered the most caffeine-concentrated drink in the largest size available (for those of you who are not Starbucks savvy, that would be *Venti*, whatever that means). I then came back to the table and sat down for a few seconds sipping my coffee before I gave Buzz the *I-don't-like-you-anymore* look. I felt violated.

"How on earth do you continue living the way you do if you thought all of this is going to happen? Why pursue any serious scientific education if someday -- soon according to you-- our intelligence is going to be grossly shadowed by machines? In fact, why work or accumulate wealth for that matter? It will all be pointless, won't it?"

Looking a bit embarrassed, Buzz almost apologetically concurred, in his own formal way.

"The arrival of AI will have a profound effect on all financial concepts including the way we invest our money."

And John had something to say about the investment subject too.

"Even if you leave AI out of it, the world rush will have serious ramifications on the way we invest money and sustain wealth in the future. Last century was full of examples where investment opportunities were lost because of investors' inability to recognize the speed of change in the markets. AT&T and Western Union were going head to head in the second half of

The Future

the nineteenth century on the telegraph and telephony fronts[9]. They made a pact to stay out of each other's way. AT&T took the telephony market and Western Union took the telegraph market[10]. I am sure members of the Western Union executive team were congratulating each other on the visionary decision that they made. And as you know, fundamental changes in our social order over the decades made AT&T one of the largest giants of the twentieth century, while Western Union has become a money transfer company overshadowed by banks and other financial institutions. Such impactful changes are about to happen much faster in the twenty first century and will leave unwary investors trapped in obsolete portfolios."

"But long term investments are safe, right? That's what I keep hearing whenever I turn the TV to CNBC."

As if they both forgot their differences and were united on the new mission of pounding me, Buzz joined John in a two-flank assault that had me in the middle. Buzz started from the right.

"Alright, tell me where would you think you should invest your money? Where is that safe haven that should be the enduring home for your well-earned money?"

"I would say real-estate is a good bet."

"If you give real-estate a serious long-term thought, it might not be as appealing as you might think. For starters, we might be able to do much more with less space. We are quickly miniaturizing everything around us and efficiently embedding

---

[9] "History::AT&T". Cybertelecom. http://www.cybertelecom.org/notes/att.htm.

[10] In fact, at one point in time -- to be exact in 1877 -- Bell offered to sell their telephony Patents to Western Union for $100,000. The offer was refused.

things in all types of spaces. In the future, your medical doctor will never need to physically see you. He will probably get a dump of everything he needs about your vital signs from the devices implanted in your system, including the gene sequence of the mutant cells brewing in your intestinal track, the complete continuous read of your blood sugar levels taken from multiple veins in your body simultaneously, and your blood pressure readings anomalies for the past two years, and he will get all of this from the comfort of his home office. Clinics won't be needed, and that is but one of many other services that will require less space, from restaurants, to department stores, to furniture warehouses. Advances in technology will make it feasible for people to shop for all retail products online, probably through some sophisticated virtual reality experience."

And before I had the chance to digest all of this, John launched his argument from the left.

"Furthermore, most investments in land are in urban areas. Price of a piece of land in downtown New York might be a factor of 1,000,000 compared to the price of a piece of arid land in the Sahara Desert. Even in comparison with less hostile environments, urban land can be a factor of 1,000 more expensive than rural. This might change in the future. The urge to be close to downtown might go away. Transportation might become so fast and efficient, that a hundred miles commute might be doable in 10 minutes. Moreover, the need to commute itself might disappear altogether; virtual reality experiences might become complete enough to eliminate the requirement for anyone to be in the office, or, for that matter, for any face-to-face communications. Cisco offers a product called Musion that allows holographic representations of remote objects. You might have seen a demo of that on CNN during the 2008 elections. Seamless virtual face-to-face meetings might happen with groups sitting in different continents. In short, urban land might soon lose its appeal."

"But I will always need a house and house values should climb steadily, like they have been doing for centuries, right?"

"You'll have to start thinking like a twenty first century investor otherwise you won't survive the coming Tsunami. House prices might be riskier. In most places, the price of the house is at least three times the price of the land where it was erected and therefore the bulk of the house value is not in the land where it was built, but in the structure itself. Now, with the fast advances in manufacturing and construction technologies that took place during the past few decades, what you put in your house might become obsolete very fast, necessitating renovation. In the future, you potentially could modernize your whole house in two weeks. It already happens in rooms like kitchens or garages. Owning a house might become as depreciative a mechanism as owning a couch or a car; something to be replaced every few years and therefore an unattractive investment vehicle, if you can even call it that in the future."

"John, buddy, what are you saying here? Where do you expect me to invest my money in the future?"

"You invest in what people want to buy. As simple of an answer as this might sound, there is much more to read in it. The statement *what people want to buy* has never been so fluid, throughout history. Although the *what* has always changed with time, it has never changed this fast. In the past two decades, the amount of money that moved from the brick-and-mortar world into the digital world has been flabbergasting. A serious paradigm shift in the way we invest has happened in less than two decades. If you follow the money, you'll find that it has been moving at incredible speeds. Basically, in the future, the spending habits will change so quickly from one product to another and from one market to another that the concept of long term investment will disappear. And it is unlikely that this trend will slow down."

"I don't know about that. What could be invented in the future that would make me dole out my money on, instead of buying a house?"

John went into what seemed to be a deep thought process for a few seconds and then he put on the table something surprisingly plausible.

"Alright, this is from the top of my head. Imagine that we start creating life-impacting treatments to debilitating or frail-inducing conditions. Won't you be willing to pay for medical services that energize and revive your health and restore your youth, as much as you will on a house mortgage, if not more? I am sure that if you sit down for a couple of hours you can come up with a dozen more examples."

To listen to the discussion at hand, you would not believe that these two were at each other's throat only a few minutes earlier. I mean the last statement was something that you'd expect to hear from Buzz (and you would brace yourself from John's retaliation). I was stoned. The foes had suddenly become comrades against the evil force represented in me. I wanted to respond, but I was completely drained, so I nodded slightly in

The Future

agreement and let him finish raging his war on my future financial plans.

"The point is that the investment vehicles of the future will change much faster with time, making it hard for anyone to really feel financial security. Cash might be king, you say. True, but you'll have to make it first, and you'll have to protect it against the fast moving economic tornado."

Then helping conclude the assault, Buzz sneaked his last bit which took me over the top.

"That is all assuming that fortunes and investments might make sense in the future. An alternative path is with artificial intelligence. When such machines exist, we won't be thinking about wealth anymore. In fact, money itself will probably have no value and models of living will be created and practiced that are much different from the ones we have now. To extrapolate on that path will be speculation."

That was about all I could take on a Saturday morning. I don't mind enjoying a nice chitchat after work and only if it ends on a positive note, but this early in the day, on the weekend and on such a depressing subject, was too much for my taste. I was about to plan my quiet, slow disengagement from the discussion so I could bolt out and head home when the technology, which I completely despised at that instant, came to the rescue; my cell phone rang. I picked it up and my wife was on the line wondering what happened to me. I told her that I went out to get some coffee and was coming back home.

I then turned to the *united-we-stand* duo and pointed my index finger to the phone explaining that the wife needs me. I also gave Buzz one more *what's wrong with you* look and wondered how he slept at night when he thought this mindboggling transformation of our way of living is eminent.

Then I went to the counter and grabbed a bag of *Slightly Sweetened Guatemalan Casi Cielo – Medium Intensity* coffee bag and headed home. On the way back, I realized that the trip to Starbucks was useless; I felt more tired than when I came in.

The Future

After Thoughts

The concept of runaway AI is pivotal to the singularity and an integral part of the model for extraordinary machine capability growth. It is assumed that an intelligent machine will be able to create a more intelligent machine, and then the paradigm of *"machine improving itself"* should yield ever growing artificial intelligence that could theoretically become infinite. In this section, we will discuss why such logic might be grossly exaggerated.

Let's address the mathematical order first, and then we will backtrack into physics limitations. Assume that the intelligence level of the first true artificial intelligence machine, M_1, has the normalized value 1. The newly created intelligent machine, M_2, will have the intelligence score $1+a_1$, and will be able to create M_2, the next generation intelligent machine, that posses the ever increased intelligence $1+a_1+a_2$. And so forth. Eventually, the most intelligent machine will attain the intelligence level:

Intelligence of Machine $n = 1 + \sum_{k=1}^{n} a_k$

Given that $a_k > 0$ for whatever k, it might seem that such a series is divergent, i.e. it climbs to infinity. However, that is not the case. By definition, a series is convergent if there exists a number C, such that for some $\epsilon > 0$, there exists an integer N such that for all $k > N$:

$$|S_k - C| < \varepsilon$$

That is the convoluted way of saying that a series is convergent if its sum is not infinite. There are multiple ways to prove that a series converges, one of them is to apply the convergence test, which states that a series is convergent if:

$$\lim_{k \to \infty} a_{k+1}/a_k < 1$$

An example of such convergent series is:

$$1 + \frac{1}{2} + \frac{1}{4} + \frac{1}{8} + \frac{1}{16} + \frac{1}{32} + \cdots$$

because

$$\lim_{k \to \infty} a_{k+1}/a_k = \frac{1/2^{k+1}}{1/2^k} = 1/2$$

If intelligence growth of a machine follows the above series, the growth of intelligence will not spiral out of control. In fact, the series above is a geometric series of common ratio ½, the sum of which will be:

$$S = \frac{1}{1 - \frac{1}{2}} = 2$$

As hard as the series pushes, it will converge to a total sum of two, twice the first term. If runaway AI grows according to this series, it won't run that far; the most intelligent machine obtained after an infinite number of iterations in intelligence growth, will be twice as intelligent as that of the first machine. That is by no stretch of imagination a singularity inducing event.

From physics point of view, we know that there is no such thing as infinity. No matter how fast intelligence growth, it has to converge to a finite number. Just like all processes that start high, intelligence growth of this type will need to saturate. This rule applied to so many physical models in the past. Transistor capacity, Internet adoption, and market capitalization of Google

The Future

are all examples of what seemingly started like runaway processes but concluded into bound states, limited by the available resources, be it physical packing density, population of the globe, or the depth of pockets for average investors. The same thing will happen to AI when it arrives. At some time, it is going to become very hard for the next generation machine to improve more than a very small incremental percentage, and saturation will take place. The intelligence growth series will converge. How high a number will it converge at? There is one way to answer that question: Wait for it.

Science and Religion: To Be Continued
This discussion is never complete without a trek through the religious debates. Although science is constantly associated with atheism and more often than not science and religion seem to be sitting on opposite sides of the fence, I am a strong believer that our world was not built by accident. It's as if the laws of physics were set in place by some programming hand and a push was given here and there every now and then to make the universe we live in run what it's programmed to run. When it comes to taking sides, and for the record, I am a believer.

This subject is so wide and the debates are so intense, it warrants another book. So, when the new edition arrives and discussions on religion heat up, you'll know which side I am on.

ENDNOTE

When I first contemplated publishing my thoughts, I tried to write a book that reflects my ideologies, philosophies and beliefs, but then I realized how prejudiced that would be, especially given the contentious nature of the subjects discussed. With time, it became evident that the fair thing to do is to lay both sides of the argument on the table and leave it to the reader to formulate his own opinion. Of course, I am fully aware that my predispositions must have involuntarily crawled right between the lines, but, for my sake, I hope it's so subtle[1] that you can't even notice it!

Looking back at the book, I can see that it evolved around two main subjects: the *future* and *longevity*. For the longest time, both topics belonged to the realm of fiction and were never taken seriously, and until recently only a few brave souls tackled these risky subjects that were unmistakably in the realm of

---

[1] Science is probably the most beautiful artwork performed by man. Its completeness and beauty are reflected in the fact that only the most competent, most brilliant, and most diligent deserves the title scientist

Endnote

witchery and soothsaying. Then came the pioneers, like Ray Kurzweil and Aubrey de Grey, who pretty much legitimized both fields and brought them center stage.

The first part of the book was about the quest for youthful long life. The fundamental debate is about the existence of a cap on the length of human life. Although evolutionary biologist will argue strongly against the possibility of life extension, the basic fact remains that scientists have been able to alter, increase to be more specific, the age of small animals in the lab. There is no evolutionary reason to believe that such advantageous characteristics apply selectively to roundworms, fruit flies, and mice, but not to humans. If we can make mice live longer, we should be able to do the same for humans. And the proof of concept is already in the making. For example, it is a known fact that practicing *calorie restriction diet* is one way to extend life. Eat 30% to 40% less calories than the normal diet, steadily and consistently, and you'll live healthier. The order is very tall though. A normal diet in itself is a tough one to follow. 30% to 40% less calories than that, and you are at the always- famished state. If you practice calorie restriction, not only you won't be able to have even one good, filling, satisfying meal, you'll have to stay hungry all the time. That's hardly the desirable extension of time I had in mind. It's as desirable as wanting to extend flight delays while waiting for your flight on thanksgiving night. The alternative is drugs. It's not as bad as it sounds. Calorie restriction diets can be simulated by activating a gene in the body called Sirtuin. Take the drug and eat as you wish (within reason) and you can still get the benefits of calorie restriction. At least this is the promise and the hope of the trials on life extension drugs like Resveratrol.

No one can argue the fact that the merger of technology and medicine has created powerful tools that have transformed the medical field. The exponential growth in technological power will definitely translate into exponential growth in combating diseases. Will we be able to translate that into quality life extension paradigms? Unfortunately some of these answers might have to wait a lifetime and we don't have many of those in our lives.

The second part of the book was about the future. Fantastic future predictions, which have been concocted so many times through so many channels, be it books, articles, movies or TV shows, were *scientifically* addressed by Ray Kurzweil. The end-result, however, is not as gracious as I dreamed it would be, no matter how congenially Kurzweil kneads the finish line. We seem to be like a family going on vacation: the rush starts at 5:00 AM with a frantic attitude saturating the atmosphere. Bags get stuffed in a hurry, don't-forgets get thrown around nervously, and the dash to the door is madder than that of the poor souls running in front of the bulls in Pamplona. You charge furiously, and tense like a turkey on thanksgiving, only to arrive to your destination so you can...slow down and relax.

Our world rush might take us to that vacation spot, but with one catch: the vacation is pretty much permanent. If we get to do some of the wild things that the eccentric people of our time say they want to do, our lives will be turned upside down. From my prospective, the only thing that we will be able to do successfully is vacationing. Every other function will be moot. When robots become clever, they will rule. They will do everything, and I mean everything, and they will do it so right, it won't make sense for any of us to do any work, from the very mundane to the extremely intricate. In fact, our lives will be one

Endnote

big relaxing recess and everything we will do, we will do it for recreation. After all everything will be doable, an order of magnitude better, by some crap box. And if you detect a hint of hostility in my attitude, then you are not far off. Add a splash of nausea, a spoonful of dismay, and a handful of fear and you will sum up my sentiments about the whole endeavor. You see, in addition to the idle state of living that artificial intelligence will enforce on us, a machine governed future will have no hope, no ambition, and no quest for betterment. And I for one won't know what to do with my life if I did not have that.

The two fields are closely connected and the promise that they will bring to humanity if the exponential progress patterns sustain is enormous. Just like the reader, I really don't have a set position on how far will this go. Will we be able to expand our technological powers indefinitely and create the giant machine that will run our world and redefine our way of living or is this just a blimp on the timeline chart that will plateau soon? Time will tell. I am sure it will all be clear within one decade. I will be watching closely, and if all goes well, I will keep you posted.

ACKNOWLEDGEMENTS

I -- like most wannabe authors set on what seemed to be an exciting mission of writing a book-- had no idea what I was getting into. What started as a walk in the park turned into a mountain, rock climbing adventure. If it were not for the fuel, in the form of constant encouragements from friends and family, I would still be sitting around the breakfast table talking to my wife about how strenuous it was to try to write a book and why the task is left to the few that can go the distance.

First on the to-be-thanked list are the pioneers that made science happen. The models, theories and concepts that were created are more beautiful than the most exquisite art ever made. From my prospective, Shakespeare's sonnets are no rival to Archimedes's equilibrium of planes theories, Michelangelo's Sistine Chapel is no match to Newton's infinitesimal calculus and, if you want to go eccentric, Pablo Picasso lives in the shadows of Nikola Tesla.

Acknowledgements

Now on to the living. I like to thank my role model and my idol, my father, for making me what I am. I want thank the sweetest thing in the world, my mother, for being my number one supporter in my life and for believing in me more than anyone else. I owe them everything.

I want to thank my wife for reading, criticizing, disapproving, complimenting, counseling, recommending, condemning, censuring, correcting, appreciating, , and debating the material; for the support she gave me throughout the writing process; for being the good wife and the incredible mother that she is; and for making a fine cup of coffee. Given that she is a mathematician, it was hard to slip mediocre content by her, but maybe that's what got the book through to the finish line. And while we are on the family list, I like to thank Omar for getting more excited about the book than I did. At one point in time, I seriously deliberated taking a writing break before going into the last two chapters (which I am sure would have shelved the book indefinitely), but Omar filled up the fuel tank and spilled over. His encouragements were unabated. At the end, he was the one who created a book out of a Word document. The artwork, handiwork and legwork that I know I could never do well enough, was beautifully done by him. I owe him the finished product.

I also like to deeply thank Lisa for reading, editing and correcting my faults. And while I am at it, I would like to thank her for being my first American teacher. I would say English teacher, but what she taught me was much more than a language. And if you ever get to meet Lisa, don't be fooled by her blonde looks. She is a *wicked-smart* Bostonian. In fact, she is probably one of the smartest people I know.

I like to thank Walid Hassan for diligently editing the book, providing *frank* feedback, and being the supportive friend that he has always been.

Not quite last, but definitely not least, my full gratitude to my father-figure, my older brother Walid, for putting me through school, all the way through the Ph.D. program, and reshaping my life altogether.

And then there is Shadman Zafar. He was the one who set me on this track by handing me the first book I read on the subject. His unwavering belief in my abilities was the reason you are reading this book. He encouraged, suggested, criticized and believed so heartedly; so much so that sometimes I feel that his name should be on the cover too.

REFERENCES

Chapter 1: Introduction

- Michelson, A., et al. (1928). "Conference on the Michelson–Morley Experiment Held at Mount Wilson, February, 1927". Astrophysical Journal
- Lev B. Okun (1989), "The Concept of Mass". Physics Today

Chapter 2: Science Eccentricities

- Brian Greene (2003). "The Elegant Universe". Vintage. NY.
- Frank close (2004). "Particle Physics". Oxford University Press. Oxford.

- Peter WOIT (2006). "Not Even Wrong: The failure of String Theory and The Search For Unity in Physics". Basic Books. NY.

- Isaac Asimov (1966). "Understanding Physics. Barnes and Noble. NY.

- Albert Thomas Fromhold, Jr. (1981)"Quantum Mechanics for Applied Physics and Engineering". Dover. Ontario.

- Richard Feynman, Robert Leighton, Matthew Sands (1964). "The Feynman Lectures on Physics". Addison Wesley. MA.

- Lee Smolin (2006). "The Trouble With Physics". Houghton Mifflin. NY.

- Roger Penrose (2004). "The Road to Reality". Vintage. NY.

- Stephen Hawking (1988). "A Brief History of Time". Bantam Books. NY.

- David Lindely (2008). "Uncertainty : Einstein, Heisenberg, Bohr, and the Struggle for the Soul of Science". Random House. NY.

- Richard Feynman (1995). "Six Easy Pieces". Basic Books. NY.

- Stephen Hawking (2007). "The theory of Everything: The Origin and Fate of The Universe". Phoenix Books. CA.

- James Harle (2002). "Gravity: An Introduction to Einstein's General Relativity". Benjamin Cummings.

- Michael E. Peskin, Daniel V. Schroeder (1995). "An Introduction to Quantum Field Theory. Westview Press.

References

- Adams, D. J. (1980). "Cosmic x-ray astronomy" A. Hilger. Bristol.
- Arnett, D. (1996). "Supernovae and nucleosynthesis: an investigation of the history of matter, from the big bang to the present". Princeton University Press . Princeton, NJ.
- David Griffiths (1987). "Introduction to Elementary Particles". John Wiley & Sons.

Chapter 3: Robotics

- Luger, George; Stubblefield, William (2004). "Artificial Intelligence: Structures and Strategies for Complex Problem Solving (5th ed.)". Addison-Wesley.
- Nilsson, Nils (1998). "Artificial Intelligence: A New Synthesis. Morgan Kaufmann Publishers". San Fransico.
- Russell, Stuart J.; Norvig, Peter (2003). "Artificial Intelligence: A Modern Approach (2nd ed.)". Prentice Hall. NJ.
- Poole, David; Mackworth, Alan; Goebel, Randy (1998). "Computational Intelligence: A Logical Approach". Oxford University Press. NY.
- Winston, Patrick Henry (1984). "Artificial Intelligence". "Addison-Wesley. MA.
- Crevier, Daniel (1993). "AI: The Tumultuous Search for Artificial Intelligence" BasicBooks . NY.
- McCorduck, Pamela (2004). "Machines Who Think (2nd ed.)". A. K. Peters . MA
- Buchanan, Bruce G. (Winter 2005). "A (Very) Brief History of Artificial Intelligence" (PDF). AI Magazine.

- Dennett, Daniel (1991). "Consciousness Explained". The Penguin Press. NY.
- Dreyfus, Hubert (1972). "What Computers Can't Do". MIT Press. NY.
- Dreyfus, Hubert (1979). "What Computers Still Can't Do". MIT Press. NY.
- Dreyfus, Hubert; Dreyfus, Stuart (1986). "Mind over Machine: The Power of Human Intuition and Expertise in the Era of the Computer". Blackwell. Oxford, UK.
- Gladwell, Malcolm (2005). Little, Brown and Co. NY.
- Haugeland, John (1985). "Artificial Intelligence: The Very Idea". MIT Press. Cambridge, Mass.
- Hawkins, Jeff; Blakeslee, Sandra (2004). "On Intelligence". Owl Books. NY.
- Hofstadter, Douglas (1979). "Gödel, Escher, Bach: an Eternal Golden Braid". Vintage Books. NY
- Kurzweil, Ray (1999). "The Age of Spiritual Machines". Penguin Books. NY.
- Barrett, L. (2002). "Human Evolutionary Psychology". Princeton University Press. NJ.
- Drescher, G. (2006). "Good and Real: Demystifying Paradoxes from Physics to Ethics". Bradford Books.
- Kahneman, D., Slovic, P. and Tversky, A. (1982). "Judgment Under Uncertainty: Heuristics and Biases". Cambridge University Press. UK.
- Pearl, J. (2000). "Causality: Models, Reasoning, and Inference". Cambridge University Press. UK.

References

- Russell, S. and Norvig, P.(2002). "Artificial Intelligence: A Modern Approach". Prentice Hall. NJ.

Chapter 4: Bio-Universe

- Aubrey de Grey (2005) "Ending Aging: The Rejuvenation Breakthroughs That Could Reverse Human Aging in Our Lifetime". St. Martin's Press. NY.
- The Immortality Institute (2004) "The Scientific Conquest of Death". LibrosEnRed
- Philip Lee Miller, Monica Reinagel (2005) "The Life Extension Revolution: The New Science of Growing Older Without Aging". Bantam Books. NY.
- Warner, H; Anderson, J; Austad, S; Bergamini, E; Bredesen, D; Butler, R; Carnes, Ba; Clark, Bf; Cristofalo, V; Faulkner, J; Guarente, L; Harrison, De; Kirkwood, T; Lithgow, G; Martin, G; Masoro, E; Melov, S; Miller, Ra; Olshansky, Sj; Partridge, L; Pereira-Smith, O; Perls, T; Richardson, A; Smith, J; Von, Zglinicki, T; Wang, E; Wei, Jy; Williams, Tf (Nov 2005). "Science fact and the SENS agenda. What can we reasonably expect from ageing research?". http://sens.org/index.php?pagename=mj_about_mission
- Nuland, Sherwin. (February 2005). "Do You Want to Live Forever?". Technology Review.
- James D. Watson (2004) "DNA: The Secret of Life", 2004. Alfred A. Knopf. NY.
- Matt Ridley (2006) "Genome: The Autobiography of a Species in 23 Chapters". HarperCollins Publishers. NY.
- Britt, Robert Roy. March 9, 2005. "Anti-Aging Prize Tops $1 Million". LiveScience.

- Ben Best (December 2007). "Book Review: ENDING AGING". Life Extension Magazine. Life Extension Foundation.

- Kinzler, Kenneth W.; Vogelstein, Bert (2002). " The genetic basis of human cancer ". McGraw-Hill. NY.

- Henze K, Martin W (2003). "Evolutionary biology: essence of mitochondria". Nature.

- American Cancer Society (December 2007). "Report sees 7.6 million global 2007 cancer". Reuters.

- Hausen H (1991). "Viruses in human cancers". Science.

- Dingli D, Nowak MA (2006). "Cancer biology: infectious tumor cells". Nature.

- Alberts, Bruce (1994). "Molecular Biology of the Cell". Garland Publishing. NY.

- Grifiths, JF (1999). "Modern Genetic Analysis". Freeman Publishing. NY.

- Losick, Richard (1999). "Biological Sciences 10: Introduction to Molecular Biology". Harvard University, Department of Molecular and Cellular Biology. MA.

- Voet, Donald and Judith Voet (1995). "Biochemistry". John Wiley & Sons. NY.

- Bob Buchanan, Wilhelm Gruissem, and Russell Jones (2002). "Biochemistry & Molecular Biology of Plants". John Wiley & Sons. NY.

- Frank Stephenson (2003). "Calculations for Molecular Biology and Biotechnology: A Guide to Mathematics in the Laboratory". Academic Press. MO.

References

- William H. Elliott (2009). "Biochemistry and Molecular Biology". Oxford University Press. USA.

- Lauren Pecorino (2008). "Molecular Biology of Cancer: Mechanisms, Targets, and Therapeutics". Oxford University Press. USA.

- Lizabeth A. Allison (2007). "Fundamental Molecular Biology". Wiley-Blackwell.

- John Kuo (2007). "Electron Microscopy: Methods and Protocols". Humana Press. NY.

- Pete Shanks (2005). "Human Genetic Engineering: A Guide for Activists, Skeptics, and the Very Perplexed". Nation Books. NY.

- John Avise (2004)."The Hope, Hype, and Reality of Genetic Engineering: Remarkable Stories from Agriculture, Industry, Medicine, and the Environment". Oxford University Press. USA.

- Sandy B. Primrose, Richard Twyman (2006). "Principles of Gene Manipulation and Genomics". Wiley-Blackwell.

- Howard M. Fillit, Alan W. O'Connell (2001)."Drug Discovery and Development for Alzheimer's Disease". Springer Publishing Company. NY.

- George Grossberg(2009)." Alzheimers: The Latest Assessment & Treatment Strategies". Jones & Bartlett Publishers. MA.

- Jorge D. Brioni, Michael W. Decker (1997). "Pharmacological Treatment of Alzheimers Disease: Molecular and Neurobiological Foundations". Wiley-Liss.

- Nancy Smyth Templeton (2008). "Gene and Cell Therapy: Therapeutic Mechanisms and Strategies". CRC Press.

- David V. Schaffer, Weichang Zhou (2006). "Gene Therapy and Gene Delivery Systems (Advances in Biochemical Engineering / Biotechnology)". Springer Publishing. NY.

Chapter 5: The Future

- Ray Kurzweil (2005). "The Singularity is Near". Penguin Books. NY.
- Stanley Schmidt. "The Coming convergence". Prometheus. NY.
- Ray Kurzweil, Terry Grossman "Transcend: Nine Steps to Living Well Forever". Rodale Press. NY.
- Ray Kurzweil (2000). "The Age of Spiritual Machines: When Computers Exceed Human Intelligence". Penguin Group. USA.
- David Levy (2006). "Robots Unlimited". A. K. Peters: MA.
- Marvin Minsky (1985). "The Society of Mind". Simon and Schuster. NY.
- Damien Broderick (2001) "The Spike: How Our Lives Are Being Transformed by Rapidly Advancing Technologies" Forge. NY.
- Hubert , Dreyfus; Stuart Dreyfus2000 "Mind over Machine: The Power of Human Intuition and Expertise in the Era of the Computer (1 ed.)". Free Press. NY.
- Bill Bill (2000), "Why the future doesn't need us". Wired Magazine.
- Herbert Simon (1970). "The Sciences of the Artificial". The MIT Press.

References

- Noam Chomsky (1965). "Aspects of the Theory of Syntax". The MIT Press.
- Henry Markram (2006). "The Blue Brain Project". Nature Reviews Neuroscience.
- Feng-hsiung Hsu (2002). "Behind Deep Blue: Building the Computer that Defeated the World Chess Champion". Princeton University Press. NJ.
- Jeff Hawkings (2004). "On Intelligence". Times Books. NY.
- Nick Bostrom (1997). "How Long Before Superintelligence?". Oxfrord Future of Humanity Institute, Faculty of Philosophy & James Martin 21st Century School, University of Oxford.
- Eric Horvitz, Bart Selman, Margaret Boden, Craig Boutilier, Greg Cooper, Tom Dean, Tom Dietterich, Oren Etzioni, Barbara Grosz, Eric Horvitz, Toru Ishida, Sarit Kraus, Alan Mackworth, David McAllester, Sheila McIlraith, Tom Mitchell, Andrew Ng, David Parkes, Edwina Rissland, Bart Selman, Diana Spears, Peter Stone, Milind Tambe, Sebastian Thrun, Manuela Veloso, David Waltz, Michael Wellman (2009). "AAAI Presidential Panel on Long-Term AI Futures AAAI Presidential Panel on Long-Term AI Futures". Association for the Advancement of Artificial Intelligence.
- Aubrey de Grey (2008). "The singularity and the Methuselarity: similarities and differences".

ABOUT THE AUTHOR

Hammad Azzam is Collaboration Platform Director at British Telecom responsible for systems implementation of eCommerce solutions including Retail, Enterprise and Wholesale portals.

He received his Ph.D. in Electrical and Computer Engineering from Tufts University in 1992 and his undergraduate degree in Electrical Engineering from American University of Beirut in 1988.

Hammad began his career as a Senior Member of Technical Staff at GTE Labs, the research arm for GTE, and was part-responsible for the design and development of national network management systems that monitored PSTN switches, SS7 networks, and broadband connectivity. When the telecommunications market decided to turn cannibal, GTE was consumed by the Verizon giant and Hammad continued his career in Verizon, eventually to become Vice President – Information Technology, responsible for the design,

About The Author

architecture, development, deployment and maintenance of Retail online systems.

His areas of expertise include IT strategy and execution, eCommerce strategy and delivery, large scale project management, enterprise IT Systems, telecommunications OSS, broadband service fulfillment, information architecture and retrieval, Smart City design and operations, and new product development.

His areas of research include eCommerce, information retrieval, bioinformatics, general management, and programmable logic devices.

He lives with his wife Lina, and two daughters Tamara and Nadeen in Irving, Texas.

Index

32 nanometer integrated circuit, 125
Aberystwyth University, 66
absolute reference, 6
abstract thinking, 54, 55, 76
AC, 139
ACS, 109
Adenine, 90
Africa, 111
African American, 111
AI, 141
AI wakes up, 85
alarm system, 116
Alzheimer's, 102, 103, 104, 105, 110, 115
American Cancer Society, 109
amino acids, 91, 92, 98
amoeba, 89
amplified, 117
Analytical skills, 76
ancestors, 83
aquifers, 112
Artificial Illusion, if you know what I mean.", 65
artificial intelligence, 4, 53, 54, 55, 61, 66, 128, 138, 141, 153, 155, 161
Asia, 111
Asimo, 61, 74
assembler, 134, 135, 136, 137
AT&T, 149
atomic level, 118
atomic nucleus, 126
Aubrey de Grey, 13, 97, 99, 159, 168, 173
automation, 67
autonomous, 53, 61, 67
Avogadro's Number, 136
axioms, 5, 6, 25

bank account, 139
Barbie, 111
barrel, 112
base, 117
base sequence, 91
beryllium, 50
Beta Amyloid, 103
Big Bang, 10, 29, 30
Bill Gates, 120, 138
bioinformatics, 3
biological pathways, 102
biology, 4
Biology, 110
Bionic implants, 144
biotechnology labs, 3
Bio-Universe, 13
blackberry, 116
Blockbuster, 119
Blue Brain, 77, 78, 172
Blue Gene, 77
blueprint, 89
Bohr, 26, 47, 166
borderline, 42, 93, 107
brain functions.", 103
Brain scanning techniques, 143
brains, 6, 55, 62, 66, 74, 76, 104, 140
branes, 35
breakthroughs, 118
Brian Greene, 34, 35, 165
Building Managements Systems, 139
Burj Dubai, 114
Buzweil, 13
Byzantine, 147
caffeine, 148
cancer, 100, 102, 109, 169
cancer mortality, 109
cancer treatments, 109

Index

cardiovascular, 110
Carl Anderson, 127
Caspian Tigers, 83
Causality, 81
cell, 89
cell phone, 117
cerebral cortex, 78
champion beating chess programs, 67
change catalyst, 123
chemistry, 4
chess, 54, 67, 77, 104, 113
Chinese abacus, 104
cigar, 22, 23, 24, 32, 50, 51, 52, 95, 96, 98, 114, 115
Cisco, 150
CNBC, 149
CNN, 150
codon, 91
COGNISION™, 104
collector, 117
colony, 83
command center, 89
common sense, 131
computational power, 118
computer technology, 117
Computer vision, 75
conducting, 117
conservation principle, 104
console, 116
constant, 8
Contact, 84
continents, 112
countries, 112
creativity, 76
cures, 99, 102, 107, 115
Cyborgs, 141
cytoplasm, 89
Cytosine, 90
David Bohm, 47
death rate, 111
Deep Blue, 54, 67, 77, 104, 172
department stores, 150

desktop, 138
detect-and-destroy, 109
diamond, 112
dietary supplement, 107
digital world, 152
Discovery Channel, 2, 94, 115
distances, 7
DNA, 89
DNA origami, 136
double helix, 89
DS, 119
Dublin City University, 79
dude, 40, 125, 126, 131, 142
DVR, 119
early detection, 109
Earth, 5, 7, 9, 11, 48, 49, 56, 61, 76, 83, 86, 111, 113, 128, 145
eBay, 119
EE Times, 127
Einstein, 6, 7, 8, 11, 17, 26, 28, 31, 34, 38, 56, 85, 166
Elan Pharmaceuticals, 102
Elbot, 73
electron, 41
electronics game console, 119
electronics revolution, 118
elementary, 40
elements, 134
email, 116
Embryonic Stem Cell, 105
emitter, 117
Ending Aging, 13, 97, 168
energy, 104
engineer, 117
Engines of Creation', 135
environment, 113
Eric Drexler, 135
ESC, 105, 106
Evolution, 104
exact, 4
exponential, 104
extinct, 83
extrapolations, 74

Index

Facility Management Systems, 139
Fairchild Semiconductors, 118
FDA, 141
Feynman, 12, 46, 47, 165, 166
financial concepts, 148
flat screen, 116
floating city, 114
folds, 92
fountain of youth, 97
game stations, 118
garage door, 139
gene, 90
gene expression, 101
genetic engineers, 100
George W. Bush, 106
Gerald Schneider, 143
Global Policy Forum, 112
Gordon Moore, 118
gradations, 114
grandpa, 111
graviton, 5, 37, 38
gravity, 5
Guanine, 90
hamsters, 143
Hans Moravec, 54, 55
heart, 105
Heisenberg, 28, 29, 47, 52, 166
Henry Markram, 78
Herbert, 13
higher thinking, 146
holodeck, 79, 113
house, 1, 18, 76, 93, 113, 118, 119, 121, 135, 139, 140, 151, 152
housework, 140
Hubert Dreyfus, 80
Huckle, 48
human body, 91, 98, 102, 113, 136, 141
Human Body 3.0, 144
human genome, 109
Human Genome Project, 3
human hair, 133
I Am Legend, 100
IBM, 127
impressionist, 98
information database, 92
infrastructures, 114
integrated circuit, 118
Intel, 118
intellectual power, 80
Intelligent Quotients, 72
interactive digital TV, 118
interference fringes, 43
International Space Station, 2
Internet, 6, 147
investment, 148
investment vehicles, 153
IP, 139
iPhone, 3
iPod, 116
iPods, 96
irrational, 7
IT, 116
IT loop, 119
Jeddah, 114
Jody Foster, 84
John, 13
John Bardeen, 117
junk, 105
kidney, 105
kitchen, 113
Krippen, 100
laptop, 123
learning machine, 76
lecture, 99
Lee Adler, 105
Len Jelink, 127
Leptons, 41
life-impacting treatments, 152
lifespan, 97
lithography, 137
liver, 105
living room, 113
machine learning, 76

Index

machines, 144
magnesium, 50
Magnetic Resonance Imaging, 103
Mark Humphrys, 79
Mars Rover, 75
mathematics, 4
mature species, 85
medical tools, 115
medicine, 93
mental aptitude, 62
mental capabilities, 103, 147
mice, 55, 103, 159
Michelson and Morley, 7
micro-level treatments, 109
Mind uploading, 142
mining, 112
MIPS, 55
Mitochondria, 99
models, 4
molecular assembler, 135
monologue, 98, 131
moon, 5, 15, 28, 29, 49, 56
Moore, 118, 123, 124, 125, 127, 128, 132, 141, 143
Mount Everest, 114
movie, 43, 56, 84, 91, 100, 108
multicore microprocessors, 127
Musion, 150
nano, 4
nanotechnology, 4
Nanotechnology, 133, 134
nano-universe, 131
National Science Foundation, 144
natural language processing, 76
Nature, 104
navigation systems, 96
neurons, 77, 78, 105, 143
Neuroscientists, 103
neutrino, 41
neutrons, 8, 10, 40, 133
Nevada University, 78
new species, 143
New York City, 100

Newton, 5
next generation, 132
nitrogenous bases, 90, 91
NP-complete, 104
nuclear power, 11, 85, 120, 133, 134
nucleus, 29, 40, 42, 89, 91, 98, 99, 101, 128, 129, 133
Obama, 106
objects, 6
oil. Prices, 112
one million inhabitants, 114
outer space, 113
Pacific Ocean, 114
packing efficiency, 123
pagers, 96
Paralysis, 102
particle, 5
perfect, 4
permutations, 129
peta, 4
Pew Internet Project, 66
photoelectric effect, 38
photons, 38
physical phenomena, 4
plaque, 105
Playstation, 119
PM conundrum, 18, 56
Positron Emission Tomography, 103
precise, 4
pressure, 114
proline, 92
protein, 91
protein fragments, 143
protein manufacturing, 92, 98
protons, 8, 10, 40, 41, 133
PSP, 119
punctured pot, 111
QMOS, 147
Quantum computing, 126, 127
quantum entanglement, 49

Index

quantum mechanics, 29, 46, 47, 52, 127
quantum physics, 45, 46, 48, 49, 126, 128
quark, 128
quarks, 8, 10, 16, 35, 41
qubits, 130
rats, 84, 110
Ray Kurzweil, 11, 138, 159, 160, 171, 172
rays, 7
real-estate, 149
register, 129
Relativity, 6, 11, 12, 24, 25, 26, 28, 31, 32, 43, 56, 166
Robert Noyce, 118
robot, 61
robot technology, 140
Robotics, 12, 145
robotizing the world, 140
Roomba, 64, 65, 66, 140
Rothemund, 137
rural, 150
Rutledge Ellis-Behnke, 143
SAT, 72
Saudi Arabia, 114
Schrodinger, 26, 48, 49
Science Eccentricities, 12
scientific community, 80, 128
scientific curiosity, 84
script, 110
Seasteading, 114
Security cameras, 139
segregated society, 111
self-running, 136
semiconductor, 116
sequence, 90
settlements, 113
Single Photon Emission Computed Tomography, 103
singularity, 144, 155, 156, 173
sky, 113
Smart Buildings, 138

Smart Cities, 138
Smart Environment, 139
Smart Homes, 138
smart phones, 96
Smart Villages, 138
Space, 17
species, 82
speed of light, 6, 7, 9, 27, 28, 39, 50
Spielberg, 7
spintronics, 131, 132
Standard Particle Model, 41
Star Trek, 2, 13, 16, 100, 113
Star Wars, 63
Starbucks, 121, 123, 129, 148, 154
stateless, 90
Steady technological growth, 86
Stephen Hawking, 10, 30, 166
strand, 90
streams, 5
String Theory, 32, 33, 34, 35, 36, 165
stubborn diseases, 102
sub-human, 128
sunlight, 112
superior intelligence, 83
switching element, 117
temperature fluctuations, 114
Terminator, 56, 91
The Adler Institute, 105
The Future, 13
The Large Hadron Collider, 39
therapy, 105
thermodynamics, 104
thumb drive, 116
Thymine, 90
tides, 5
Time, 16
transformation, 154
transistor, 117, 118, 123, 124, 126, 128
trials, 103, 110, 159
Tsunami, 151

Index

tunnel, 99
TV, 116
twentieth century, 6
typewriter, 117
ultra-intelligent machine, 84
Uncertainty Principle, 28
underground, 113
underwater, 113
unexplainable phenomena, 6
universe, 29
unwinding, 90
Uracil, 91, 98
urban areas, 150
valve, 117
Very Large Scale Integration, 123
video conferences, 96
video recording, 116
virtual reality, 113
viruses, 100, 101
Voice Recognition, 74
Walter Brattain, 117
Washington Post, 112
water, 112, 121, 134, 136, 139
western civilizations, 128
Western Union, 149
What is Life', 49
Wii, 119
Will Smith, 100
William Shockley, 117
Wright brothers, 62
Xbox, 119
X-Seed 4000, 114
your blood sugar, 150
zoo, 83